USING AND APPLYING MATHEMATICS
Strategies for teachers

Compiled from contributions by:

Jeannie Billington: Noel Fowler: Jon MacKernan

Jim Smith: Jude Stratton: Anne Watson

Using and Applying Mathematics, as represented in Attainment Targets 1 and 9 and the associated programmes of study, should stretch across all other work in mathematics, providing both the means to, and the rationale for, the progressive development of knowledge, skills and understanding in mathematics.

This is a major undertaking for schools, and perhaps the single most significant challenge for the teaching of mathematics.

Non-Statutory Guidance for Mathematics
Paragraph D.1.5

CONTENTS

USING THIS BOOK

PART ONE: IDEAS TO CONTEMPLATE

    A.   OVERVIEW

              The Nature of Mathematics

               'Using and Applying Mathematics'

               The First Dimension - The Use and the Application of Mathematics

               The Second Dimension - Working in a Mathematical Way

    B   PARTICULAR ISSUES

               Some Problems with Problem-Solving

               Pointers to Open-Ended Problem-Solving

               Questions to Ask:

                      Management
                      Organisation
                      Groups v. Individuals
                      Level of Work; Success v. Failure
                      Investigating the Mathematics
                      Presentation
                      Assessment

               Are Investigations Worth While?

               Valuing Children

               The Importance of Discussion

Two Activities to Encourage Discussion

Learning Languages

C. ACTUAL LESSONS

A Series of Lessons

A Lesson Plan Network

PART TWO: IDEAS TO USE

A IDEAS ON USING MATHEMATICS

Selecting materials and mathematics appropriate for a particular task

Planning and working methodically

Checking for sufficient information

Reviewing progress at appropriate stages

Checking that results are sensible

Using trial and improvement methods

Trying alternative strategies

Completing a task

Presenting alternative solutions

B IDEAS ON COMMUNICATING MATHEMATICS

Making sense of a task

Interpreting mathematical information

Talking about work in progress and asking questions

Explaining and recording work systematically

Presenting results in an intelligible way to others

C   IDEAS ON DEVELOPING ARGUMENT AND PROOF

Asking the question "What if ....?"

Making and testing predictions

Making and testing statements

Generalising, making and testing hypotheses

Following arguments and reasoning, and checking for validity

Conjecturing, defining, proving and disproving.

Uses and Applications

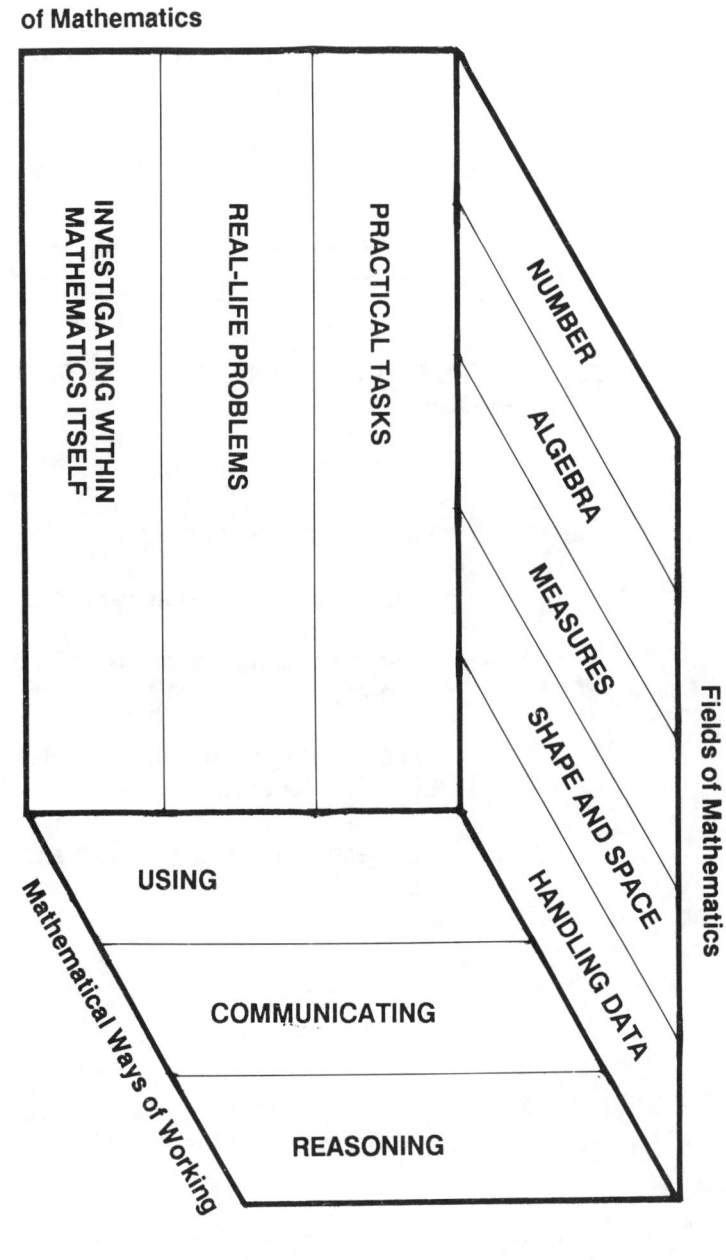

of Mathematics

- PRACTICAL TASKS
- REAL-LIFE PROBLEMS
- INVESTIGATING WITHIN MATHEMATICS ITSELF

Fields of Mathematics

- NUMBER
- ALGEBRA
- MEASURES
- SHAPE AND SPACE
- HANDLING DATA

Mathematical Ways of Working

- USING
- COMMUNICATING
- REASONING

USING THIS BOOK

This booklet is designed to give some practical help to teachers who are preparing work on Using and Applying Mathematics.

Part One is in three sections.

The first section examines the various strands concerned with Using and Applying Mathematics as laid out in the National Curriculum programmes of study, attempting to clarify the complex interrelationships. The three charts provided with the booklet belong here: they seek to clarify in tabular form the strands and the progression involved.

The second section looks at various issues that are liable to arise in implementing these strands of the programmes of study in the classroom. It raises questions, highlights potential problems and offers advice and suggestions.

The shorter third section contains an example of a lesson plan network, and an account of a complete series of lessons.

Part Two considers each of the twenty phrases listed in Non-Statutory Guidance D.1.4, and, for each of these, takes the form of a collection of short ideas, discussion points and activities, for you, for colleagues and for children.

These lists are by no means exhaustive and you should be able to add many ideas of your own. We hope that those produced here will be seen as a start: room has been left for you to add to them, and we hope that you will do just this.

Not all of the ideas described here are original, and you will no doubt have met some of them before. However, we believe that the collection will be both stimulating and useful.

Many of the ideas are designed for immediate use in the staffroom or in the classroom, but you may find others less accessible and a few you may be unable at first sight to see how to use easily or even how to use at all! This should not be a bad thing; it is intended that all the items can, and should, be worked on, and particularly with regard to the less transparent suggestions, that you will discuss them with colleagues and return to them at a later date.

The presentation of the ideas in lists is simply for convenience of reference and in order to break up what would otherwise be a monotonous stretch of print. It is in no way our intention that the skills and processes concerned should be seen as fitting into separate compartments, nor that they should be tackled individually or away from a problem-solving or investigational environment. Most, if not all, of the ideas could have been placed under several headings, and, in order to emphasise this, we have not hesitated to use a few of the ideas more than once. We hope that the fuller description of a complete piece of work given in Part One will also help to redress any imbalance.

The ideas are not age-related. The contributors to the booklet are working with a wide variety of people, and they have tried to write for teachers of the whole 5 to 16 range, as well as for initial trainers and for inservice providers. As a result, we hope that you will be able to relate the ideas without difficulty to your own learners, whether they be very young or almost adult.

For convenience the word 'children' has been used throughout to represent the learners.

It should be noted that this booklet is about the Using and Applying Mathematics programmes of study, i.e. about doing and learning; it is not about AT 1 and 9, which are about assessment. Moreover, neither this booklet, nor the programmes of study themselves, contain or refer to everything that is appropriate in this area.

PART ONE

IDEAS TO CONTEMPLATE

A. OVERVIEW

REAL WORLD

Viewing   Making Sense

Practical Tasks

Problem Solving

RELEVANCE & UTILITY OF MATHEMATICS
(Concrete)

MATHEMATICS
(Abstract)

Investigating within Mathematics Itself

Creating   Exploring

NEW IMAGINATIVE WORLDS

THE NATURE OF MATHEMATICS

Mathematics is a way of viewing and making sense of the real world. It is also the material and the means for creating new imaginative worlds to explore.

In short: mathematics is applicable to the concrete, but mathematics itself is abstract.

It is essential to recognise these two aspects:

- Mathematics is relevant to and is useful in the real world around us; it can be found in practical tasks, and it can be applied in order to tackle real-life problems.

- Mathematics can be used to explore and to investigate within itself, thereby creating new mathematics.

With this in mind, it is clear that the rationale for the over-riding importance of using and applying mathematics throughout the National Curriculum lies in the very essence of mathematics itself.

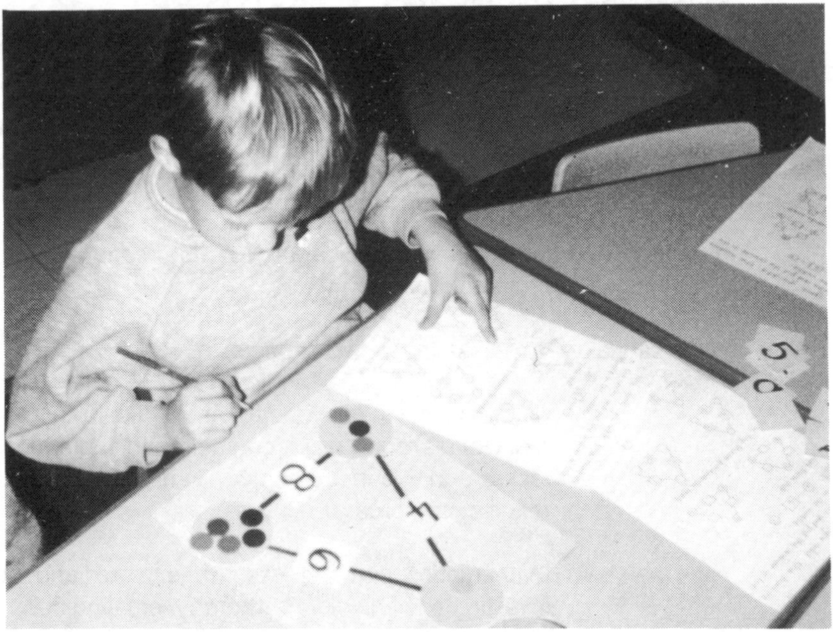

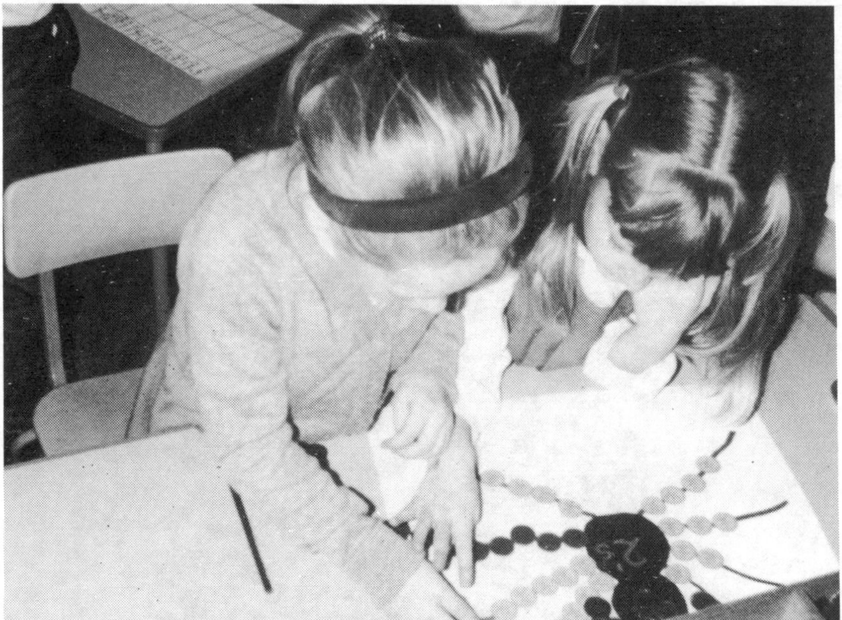

'USING AND APPLYING MATHEMATICS'

There are two dimensions to Using and Applying Mathematics as it is at present laid out in the Statutory Orders for Mathematics.

The first dimension involves the mode in which an activity is presented or is pursued. It also relates to the purpose for which the mathematics is being employed. It truly concerns the use and the application of mathematics.

The second dimension involves the mathematical skills, processes and strategies that may need to be used. It concerns the mathematical way of working.

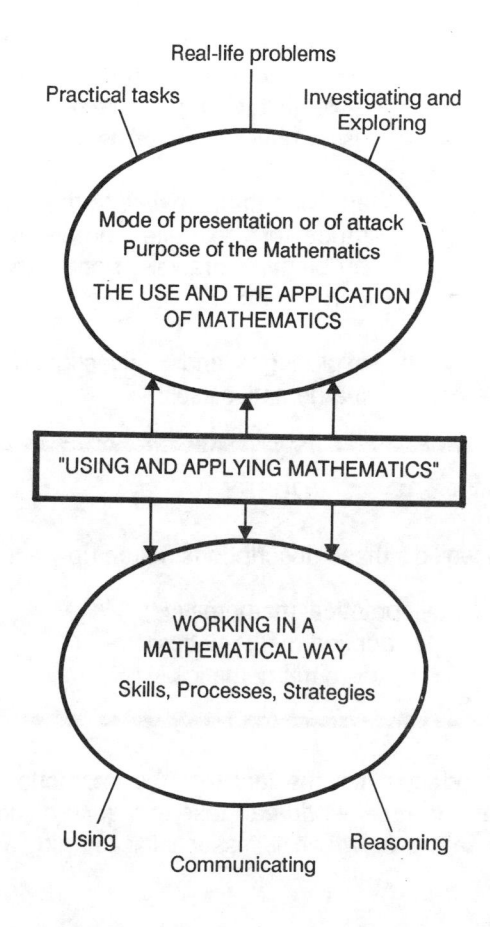

```
           Real-life problems
Practical tasks      Investigating and
                         Exploring
       ┌─────────────────────┐
      (  Mode of presentation )
      (  or of attack         )
      (  Purpose of the       )
      (  Mathematics          )
      (  THE USE AND THE      )
      (  APPLICATION OF       )
      (  MATHEMATICS          )
       └─────────────────────┘
            ↑↓ ↑↓ ↑↓
    ┌─────────────────────────────┐
    │ "USING AND APPLYING         │
    │       MATHEMATICS"          │
    └─────────────────────────────┘
            ↑↓ ↑↓ ↑↓
       ┌─────────────────────┐
      (  WORKING IN A        )
      (  MATHEMATICAL WAY    )
      (  Skills, Processes,  )
      (  Strategies          )
       └─────────────────────┘
        Using        Reasoning
           Communicating
```

THE FIRST DIMENSION

The Use and the Application of Mathematics

The Statutory Orders, in their description of AT 1 and 9, outline three modes in which an activity may be presented or tackled:

Pupils should use their mathematics

- in practical tasks
- in real-life problems
- to investigate within mathematics itself.

These modes are amplified in the Non-Statutory Guidance to form the principle that pupils should be:

- acquiring knowledge, skills and understanding through practical work, through tackling problems, and through using physical materials;

- applying mathematics to the solution of a range of 'real life' problems, and to problems drawn from the whole curriculum;

- exploring and investigating within mathematics itself.

QUERY

How well do these descriptions match up with:

practical mathematics;
applied mathematics;
pure mathematics.

The first two modes reflect the fact that the interaction between mathematics and the real world can take two quite distinct forms, involvement in each of which is essential if learning is to take place:

- Mathematics can be drawn out of the child's experiences, whether these take place in the outside world, or are, for example, to do with using structural apparatus in the classroom.

    The real world can be 'mathematised'.

- Mathematics that has been learnt can be used for real-world purposes, and in particular in the child's own everyday activities.

    Mathematics can be 'applied'.

Too great a concentration on the second of these modes can lead one into teaching, or at least to concentrating on, the mathematics that is believed to be "best" or "most useful".

Now, this requires a subjective choice which will almost inevitably be based on the history of mathematics and of the mathematics curriculum - as we see it; and hence it will be culture based.

On the other hand, by mathematising the child's own experiences, we avoid this culturism.

Moreover, since the mathematics will now come from context, we also avoid the problem of decontextualisation, which is another inevitable consequence of simply teaching mathematics in order to apply it.

The principle distinction between the first two modes, with their real world associations, and the third mode, with its embodiment in the mathematical world, can be seen in the words of two of our poets.

Wordsworth appealed to the first two modes when he wrote:

> *I've measured it from side to side:*
> *'Tis three feet long, and two feet wide.*

Whilst William Blake sympathised with the third mode when he penned the lines:

> *I must Create a System or be enslav'd by another Man's.*
> *I will not Reason & Compare: my business is to Create.*

More prosaically, the Mathematics Working Group wrote:

> *Mathematics is not only taught because it is useful. It should also be a source of delight and wonder, offering pupils intellectual excitement, for example, in the discovery of relationships, the pursuit of rigour and the achievement of elegant solutions. Pupils should also appreciate the essential creativity of mathematics; it is a live subject which is continually evolving.*

Thus for pupils to develop mathematically with understanding and feeling, it is essential that they are given access to all of:

- mathematisation;
- the application of mathematics
- exploration and investigation within mathematics itself.

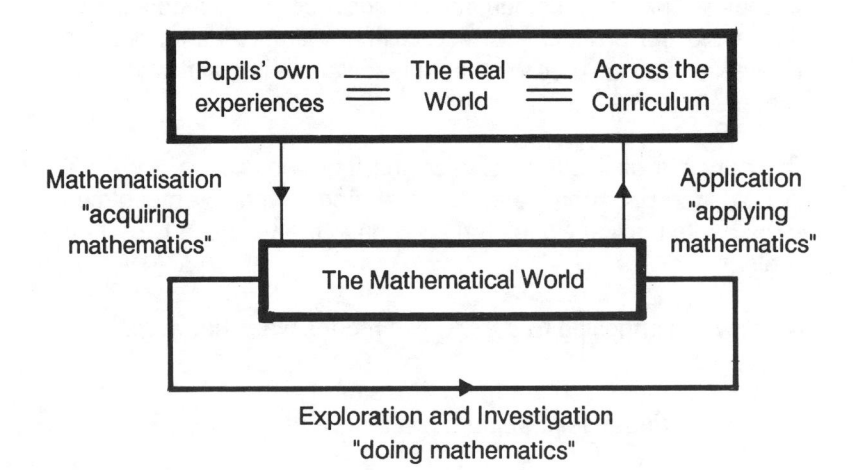

```
        Pupils' own  ==  The Real  ==  Across the
        experiences      World         Curriculum

Mathematisation                              Application
 "acquiring                                    "applying
 mathematics"                                mathematics"

                  The Mathematical World

              Exploration and Investigation
                   "doing mathematics"
```

Note that new mathematics can be derived from known mathematics, through exploration and investigation, or it can be derived through mathematisation. It cannot be derived through application, for application has to proceed from known mathematics.

It is, however, a common occurrence that, in order to apply mathematics to solve a particular problem, new mathematics has to be invented, and this need for the new mathematics can be used as an incentive to master it.

THE SECOND DIMENSION

Working In A Mathematical Way

Some of the skills, processes and strategies that are involved in learning to do mathematics were identified by the Mathematics Working Group in their Interim Report:

- doing practical work in everyday life, making models, carrying out surveys;

- listening to explanations, instructions, questions, answers;

- talking about our work, explaining what we have done, talking through our difficulties;

- thinking things through and discussing mathematical ideas and problems;

- reading mathematics from textbooks, topic and reference books, from a computer monitor;

- writing answers to questions, project reports, recording the results of discussions and surveys;

- observing what is happening, looking for pattern and relationships;

- practising skills which are important;

- using mathematics in context, in work in science, art, geography, in projects;

- relating something new to what has already been learnt;

- investigating, solving problems;

- creating mathematical models;

- trying different methods, using standard methods, using our own methods;

- trial and error, testing things out;

- using various resources, pencil and paper, materials, games, puzzles, calculators, microcomputers;

Whilst the following targets were seen as especially relevant to the development of a problem-solving ability:

- Analyse non-routine problems and use appropriate combinations of mathematical ideas and skills to obtain and evaluate solutions;

- Select, adapt and devise mathematical models of practical situations;

- Recognise patterns and structures in mathematical situations and form generalisations;

- Detect, analyse and correct errors in mathematical procedures;

- Explain the procedures used and the reasons behind them.

In the Statutory Orders these abilities are seen as falling into three overlapping areas:

- using mathematics;
- communicating mathematics;
- developing ideas of argument and proof.

These three themes can be very clearly seen in the programmes of study pertaining to Using and Applying Mathematics. Indeed, within each level, statements about using almost always precede statements about communication, which in turn precede statements about reasoning.

The themes are not simple, and each is composed of several intertwining strands. The three charts accompanying this booklet identify nineteen such strands, and show how the Statutory Orders attempt to map them through the ten attainment levels.

More recently the Non-Statutory Guidance for Mathematics attempts to combine these ideas under twenty umbrella phrases. It is these twenty phrases that are the basis for Part Two of this booklet.

USING MATHEMATICS

- Selecting materials and mathematics appropriate for a particular task
- Planning and working methodically
- Checking for sufficient information
- Reviewing progress at appropriate stages
- Checking that results are sensible
- Using trial and improvement methods
- Trying alternative strategies
- Completing a task
- Presenting alternative solutions

COMMUNICATING IN MATHEMATICS

- Making sense of a task
- Interpreting mathematical information
- Talking about work in progress and asking questions
- Explaining[1] and recording work systematically
- Presenting results in an intelligible way to others

DEVELOPING IDEAS OF ARGUMENT AND PROOF

- Asking the question "What if ...?"
- Making and testing predictions
- Making and testing statements
- Generalising, making and testing hypotheses
- Following arguments and reasoning, and checking for validity
- Conjecturing, defining, proving and disproving.

---

1   *In the June 1989 issue of Non-Statutory Guidance, the word 'Explaining' is misprinted as 'Exploring'. See the programme of study and the attainment targets at Level 3 for the correct wording.*

PART ONE

IDEAS TO CONTEMPLATE

B. PARTICULAR ISSUES

SOME PROBLEMS WITH PROBLEM SOLVING

Any application of mathematics to solve a real-life problem requires the pupil to engage in mathematical modelling, i.e. to use a piece of mathematics to model the real world situation.

It involves deciding on the mathematics to use, and of ensuring that the results that the mathematics throws up are meaningful.

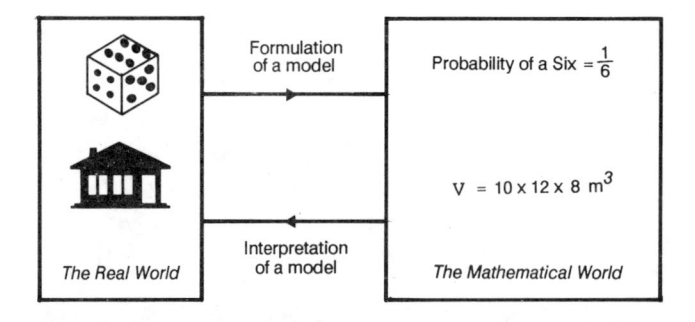

```
┌─────────────────┐   Formulation    ┌─────────────────────────┐
│   [dice image]  │   of a model     │  Probability of a Six = 1/6 │
│                 │  ──────────────► │                         │
│                 │                  │                         │
│  [house image]  │                  │  V = 10 x 12 x 8 m³     │
│                 │  ◄────────────── │                         │
│                 │   Interpretation │                         │
│  The Real World │   of a model     │  The Mathematical World │
└─────────────────┘                  └─────────────────────────┘
```

Mathematical modelling is not a natural thing to do, nor an easy thing for children to learn to do.

The problem with problem-solving is that the real world is extremely complex. It tends to yield vast, but often deficient, quantities of data and to produce all sorts of variables, unquantifiable effects and caveats.

Nor are most problems clear or precise. As a result, the most difficult thing, as in most aspects of education, is to discover the wood for the trees.

Again, it is not always clear whether mathematics can indeed help to solve a problem, or, if it can, whether the mathematics to be used should be elementary or advanced.

Consequently problem-solving, like investigating, is a very difficult skill to teach. Some might argue that neither can be taught, only learnt; but even if one goes as far as this, it is clear that schools must give their pupils opportunities for engaging in and for gaining experiences of these modes of working.

Observe that both practical tasks and real-life problems are likely to produce unequal opportunities unless care is taken to avoid this.

(As noted above, even one's perception of mathematics itself can be embedded in cultural roots and biases.)

We must continually ask whether the tasks and problems we offer avoid sexism, racism, classism, ageism, culturism, and discrimination against the disabled, the handicapped and those with exceptional needs.

POINTERS TOWARDS THE DEVELOPMENT OF SUCCESSFUL OPEN-ENDED PROBLEM-SOLVING

Giving A Choice Where Possible

This may mean a choice of several different problems, or flexibility of choice within one problem.

Presenting The Problems Verbally, Giving Maximum Visual Support Where Appropriate

At least initially: once children have gained confidence they can be encouraged to use problem-solving books or to create their own problems.

Enabling Children To Own The Problem

Encourage children to formulate problems in their own work; suggest that they pursue those aspects that interest them most; allow them to simplify, to extend or to modify problems if they wish.

Encouraging Children To Work Together, Sharing Their Ideas For Tackling The Problem

They should be encouraged to work together. It is not "cheating" to work in collaboration.

Allowing Time And Space For Collaboration And Consultation

This will vary with different children. It may mean a day, a week, or a month.

Intervening, When Asked, In Such A Way As To Develop Autonomy And Independence

Children have their own strategies for tackling problems. We should tune into these, using them to encourage the children to develop their confidence and independence. Building on mistakes and persevering when stuck are both honourable and very necessary.

Encouraging Children To Report The Progress They Are Making

This may involve reporting back to a small group, to the class or to the teacher. It may also involve sharing their working notes and diagrams.

Working Alongside The Children, Setting An Example Yourself

This might involve making your own rough notes or diagrams, discussing mathematical ideas, and in some cases offering a possible presentation.

Encouraging The Children To Present Their Work To Others

Children should be aware of the potential audience for their work. They can share their work with their peer group, the whole school during Assembly, their family, visitors, or their class teacher; but most importantly their work must be personally satisfying and something that they can return to later if they wish.

Think about the problem and + ideas

    Use your imagination

Developing a good plan

    Work alone or with someone else

Work out which corner your going to work in

    Start to collect all the tools you need

Get down to work

    Start by solving the problem and maybe you can write down what you have done already

    Put much more effert into yourself

Try more things like experiments

    Encourage your friends

Always finish it

    Try and argue you may get more ideas into your head and you may get somewhere.

    Keep on writing

When you have finely solve the problem don't stop writing

    Write your rough writing in neat

    You should now draw a digram

Be calm and concentrate

    Now be proud of yourself?

        Are you ?????

Titilayo's advice on Problem Solving

The next nine pages contain

QUESTIONS TO ASK

(How will you use them?)

MANAGEMENT

- Is there a mathematics base, with rooms and resources, which allows children easy access to equipment, stimuli, and computers?

- Is the scheme of work flexible enough to allow staff and children to follow their own plans and ideas, and to negotiate some (or all) deadlines?

- Does the school's review and assessment structure allow formative and non-intrusive assessment as an integral part of children's work?

- Is there support for mathematics as a practical subject; e.g. technicians, money for equipment?

- Can children take responsibility for some aspects of school life; e.g. displays, events, decisions?

- Are children expected to discuss work and to work together when appropriate?

- Are children allowed to decide when their work is finished?

- Have ways been devised to value work which is done in groups?

- Is there sufficient co-ordination between 'subjects' to allow children to use non-mathematics experiences to produce multi-disciplinary projects?

- Are calculators available, or are children encouraged to purchase their own?

- Should children be allowed to present work in their mother-tongue? How can this be facilitated?

- Are word-processing facilities available, especially for the children?

- Is a Mathematics Evening a viable idea?

- Can Parents' Evenings be used for mathematics displays?

ORGANISATION

How will investigational work fit into your present scheme/syllabus?

- What will you omit in order to have time for extended investigations?

Or should we re-write our schemes/syllabuses around the idea of investigational work?

- Indeed, how appropriate will schemes/syllabuses be?

Will we teach in terms of Content or in terms of Processes?

- If the latter, how do you write a syllabus in terms of processes?

How long will a typical investigation take? Hours? Days? Months?

What timetabling requirements will you want if you are to conduct extensive investigations? How long should a period of mathematical activity be? And how often?

Will you still need the same type of accommodation?

What resources will you need to purchase?

- What apparatus? Grid papers? What else?

What textbooks will you purchase?

- Or, will textbooks be inappropriate for investigational work?

What topic books will you purchase?

What teachers' resource books will you purchase?

Do you have sufficient access to calculators? To computers?

How large should a class be for investigational work?

- Will everyone in the class do investigations at the same time?

- If so, will they all do the same investigation at the same time?

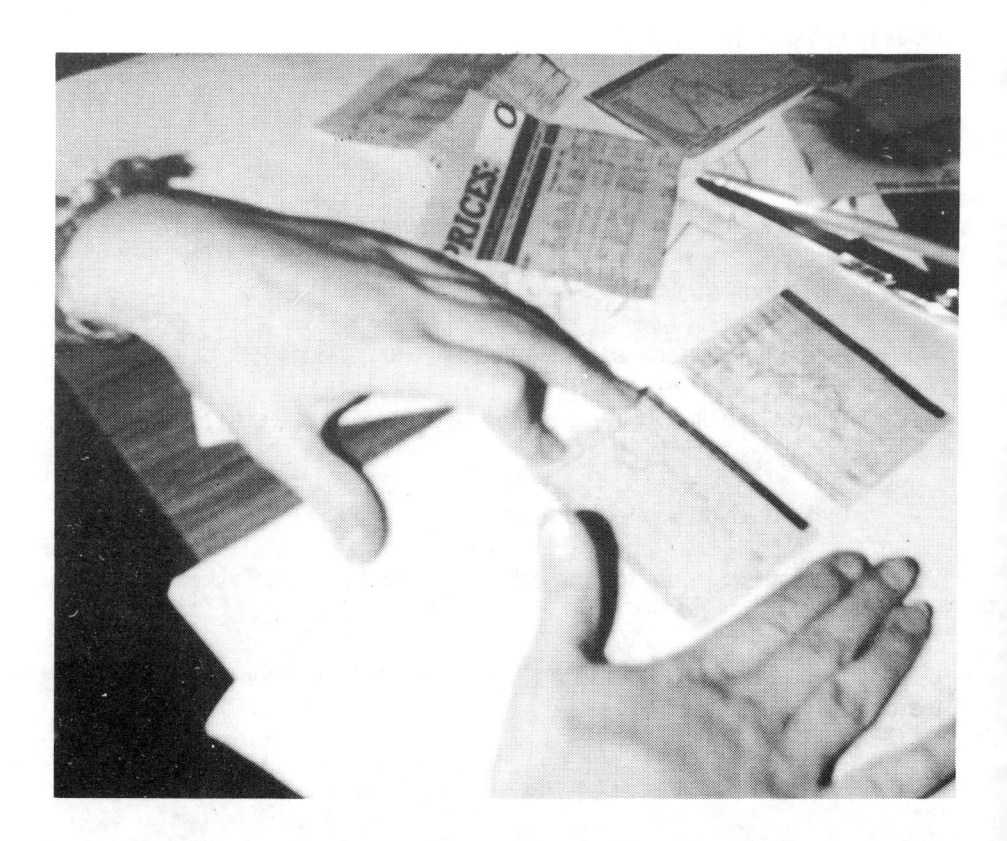

GROUPS v. INDIVIDUALS

Can one collaborate in mathematics?

What suits you personally?

What is best for children? What do they prefer? Are there differences for boys and for girls?

- Should we put them in groups? If so, how?
- Should we let them work on their own?

Which is more important: the social consequences, or the consequences for mathematical learning?

Does working in groups lead to:

- intimidation of the 'slower' members?
- interference of thought patterns?

Does it lead to faster results? And is this what we want?

Does working individually:

- give one less access to insights?
- narrow the line of development that a child is likely to undertake?
- make the exercise tedious?

Does it encourage one to internalise the mathematics more deeply?

Does working in groups tend to make one act more as a teacher than a learner?

LEVEL OF WORK; SUCCESS v. FAILURE

Many investigations can be carried out by children, and adults, of very different age and attainment levels:

- How different can these be?

- Could everyone in a school conduct the same investigation on the same day? Would this be nice? desirable? foolish?

The level of a child's attainment (surely ???) implies the level of attack:

- Does this imply the level of success?

Will the child who does 'best' on a written traditional-style test necessarily do 'best' on an investigation?

- In which case, what are the implications for banding and setting?

- And what are the implications with regard to parental expectations?

If an investigation is not solved, does this mean that it was unsuitable for the children concerned?

- Or can an unsuccessful investigation still have been a worth while activity?

- And what do we mean by 'solving' an investigation? Is the word appropriate?

Does the teacher have to be one step ahead of the children?

- Or can teacher and pupils genuinely work together towards a result?

- And how will you feel if your pupils do 'better' than you do?

INVESTIGATING THE MATHEMATICS

What do you do when you get stuck?

What do other people do when they get stuck?

What do children do when they get stuck?

What can/might children do when they get stuck?

What kinds of mathematics are useful in conducting investigations?

What kinds of mathematics are learnable through investigations?

How quickly should one introduce a calculator or a computer if number work is involved? What decides this? And who?

- Does it matter?

- Can patterns be too easily missed if one uses a computer?

- Or are they more likely to be encountered?

Is it sufficient to see the patterns involved?

- Or should one *prove* the results?

PRESENTATION

How do you record and write-up an investigation?

How should children record and write up an investigation?

- In their own way?

- Or yours?

As you carry out an investigation how messy is your work in progress?

- Is it reasonable for children's work in progress to be equally messy?

- Or should we risk discouraging them from becoming involved in the investigation through an insistence on a clear layout even for tentative work?

- At what point does a messy layout impede mathematical awareness?

- And to what extend does neatness require the use of time that might be more valuably spent?

Should a final presentation involve:
- Symbolism?
- Words?
- Essay writing?
- A display?
- Artefacts?
- Photographs?
- An oral component?
- Perhaps a cassette recording?
- Or even a video by the children?

- Should symbolism and terminology be standard? Or can it be the child's own?
- Does it matter?

ASSESSMENT

How will you assess an investigation?

Which is more important:

- The doing?

- The reasoning?

- The results?

- Or the presentation?

And which will you assess?

Which of the following are important:

Content - Processes - Proof - Communication - Approach
- Implementation - Mathematical Knowledge
- Interpretation - Attitude - Autonomy - Self Evaluation
- Ability to work in a group - Flexibility - Perseverance.

And which will you assess?

Which are appropriate for assessing an investigation:

- A detailed mark scheme?

- Impression marking?

- Ticking off Statements of Attainment without the need for marks?

- Other? What other?

Can you draw up an assessment scheme of any kind for an investigation before the investigation takes place?

- Or should you make your assessment decisions after you have seen what the children have done? - Or is this unfair?

ARE INVESTIGATIONS WORTH WHILE?

They build confidence

They develop and practise skills

They show the need for new skills

They awaken ideas

They act as lead-ins to new ideas and processes

They promote the development of new symbolisms

They promote the development of new recording techniques

They stimulate discussion

They allow children of different attainments to work side-by-side on the same problem

They stimulate motivation - Children actually enjoy them!

They encourage interactive skills, both with respect to people and materials

They encourage the use of information resources

They create valid reasons for using calculators and computers

They encourage determination and perseverance

They encourage flexibility

They encourage reflection

They encourage evaluation of one's own work

They allow the child to see their work, and mathematics in general, as their own.

They also involve proper mathematics being done properly

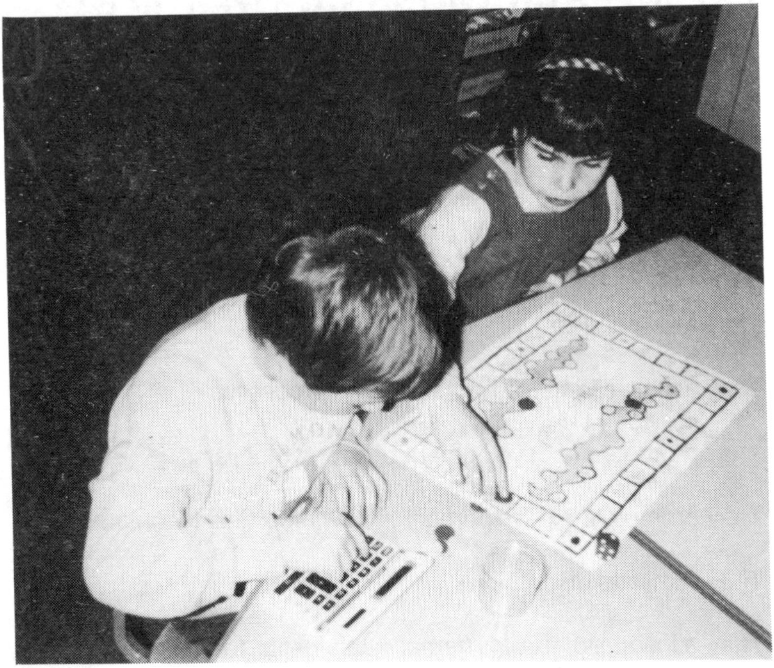

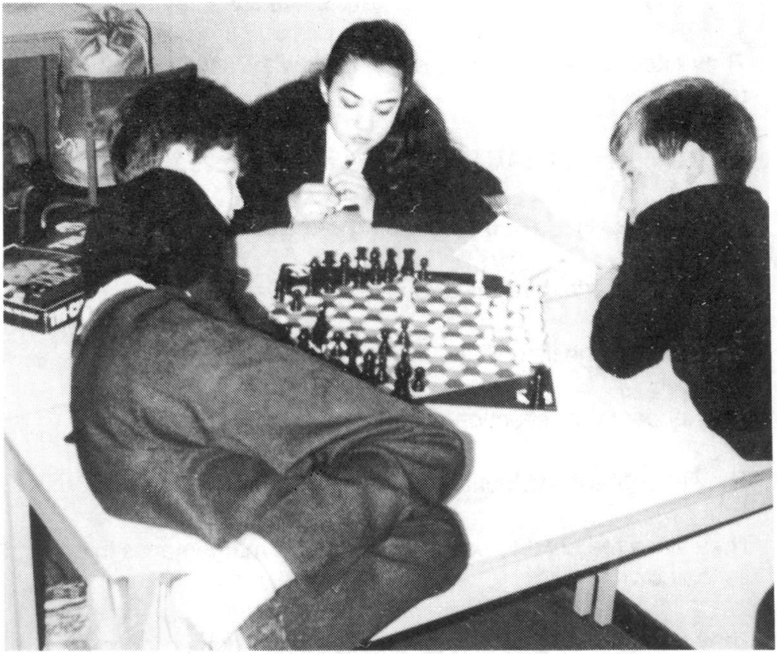

VALUING CHILDREN

Children's experiences, interests, abilities, attitudes, and cultural backgrounds are all different. This makes each child unique. It is the uniqueness of each individual child that we should learn about and value. The more we know about the uniqueness of each child, the better we can appreciate her/him and support her/his learning.

Building on children's uniqueness is vital. Once children know that their views, their interests and their experiences are not only acceptable but valued as well, they are able to develop a positive self-image and a confident attitude towards learning, and they are more likely to be considerate and responsible in a collaborative setting.

The teacher's role is crucial here. Thus, when we 'reply' to a child, it implies that we are taking the child's view seriously, even though we may wish to extend or modify that view. When we 'assess' what the child says or does, we distance ourselves from the child, and we ally ourselves with external standards which may implicitly devalue what the child has constructed.

In addition, the way we interact with children sets covert examples and expectations of how children should act among themselves when collaborating. Most children have a natural inclination to work together. We should build on this, spending time establishing successful patterns of working. This will involve:

- getting children to see us as an extra resource that they can call on when needed;

- dipping into group activities when appropriate, not as a 'judger' but also to participate and to learn from the children's discussions;

- demonstrating the value placed on learning by providing, from time to time, an audience for each group's activities.

THE IMPORTANCE OF DISCUSSION

Discussion amongst children is often undervalued, with too much consideration being paid to those occasions when grouped children have wasted their time or have failed to make progress.

However, given a real challenge and the knowledge that their views are truly valued, most children can be far more responsible learners than we often allow for. When children are encouraged to work collaboratively on mathematical activities, they discuss and share their ideas; they talk more freely and increase their language proficiency; they take greater risks in posing questions; they develop better strategies; and they support one another in their learning. Moreover, children working in this way are more likely to openly express doubts about their understanding.

This is not surprising for collaborative discussion involves:

- Initiating
    hypothesising
    making statements
    posing questions
- Eliciting
    asking for an illustration as a test of generalisation
    using evidence to challenge an assertion
    obtaining information from each other
- Extending
    extending a previous contribution
- Qualifying
    qualifying one's own or another's contribution
    repeating with modification
    providing an example
    encouraging others to continue
    answering or repeating questions
    posing a new problem

Variety of grouping is beneficial and groups should not be permanent. They need not be determined by 'attainment', nor even be decided on by the teacher - although there will inevitably be occasions when the teacher will want to ask one child to work with another for a specific reason, (e.g. to demonstrate how to use a particular computer programme or how to play a particular game), or to encourage mixed-gender groupings.

In addition to working with each other, children can spend time discussing their work with visitors and with others. This gives their work added importance and it provides an alternative audience and an additional means of consolidating their understanding.

When children are exploring a new idea, their language is often tentative and inexplicit, there are often unfinished sentences, and sometimes there is silent interaction. Now, when they share their work with others rather than work in isolation, they have a greater need to formulate and to order their ideas, and their language becomes more precise and descriptive. Working in this collaborative way they initially discuss their work in their own terms; later, as they develop some basic understanding, 'accepted' mathematical terms can be introduced; and, later still, they can come to comprehend the often impenetrable language of mathematics textbooks. These developments in language and understanding are important for children.

When new aspects of mathematics are introduced, particularly if this is done through textbooks, it is almost impossible for children simultaneously to comprehend new mathematical concepts and new sophisticated mathematical vocabulary. Indeed, when exploring new aspects of mathematics, mathematicians themselves do not invent precise mathematical symbols, notations, or terminology until they have spent some time looking at the mathematics in question. Yet, despite this, we often expect children to master the two components together.

The majority of mainstream publishers' mathematics resource books emphasise certain 'received' vocabulary (e.g. 'larger than', 'denominator', 'simplify') at the expense of other aspects of mathematics language. This can result in the teacher engaging the children in meaningless exercises from textbooks instead of in practical activities that involve them in using language in a real context.

It is important for teachers to appreciate the complexity of the language of mathematics. A proper appreciation should include an understanding of:

- symbols (e.g. '4', '+', '→')

- notation (e.g. place - value, 'xy')

- special mathematical words ('parallelogram')

- 'ordinary' words which take on a special meaning in mathematics ('straight', 'plot', 'group')

- words that aid concept development ('same', 'difference')

- words that are similar in appearance or sound ('thirteen' v. 'thirty')

Even the ideas behind the number-names can cause problems. Native speakers, as well as bilingual learners, often have great difficulty with the '-teens' and the '-tys', not just because they sound similar, but also because the teens are not consistent. Thus, "Onety-one, onety-two, onety-three, ...", or even "Oneteen, twoteen, threeteen, ..." would make more sense to children than does "Eleven, twelve, thirteen, ...". With regard to bilingual children, it should be noted that some languages, such as Chinese, have a totally logical system; on the other hand, many languages have a far more sophisticated system of number-names than our own.

In addition to this, the language of mathematics textbooks raises the issue of complex linguistic constructions, especially with regard to the statements, including those of a conditional nature, and the commands which permeate textbooks and many teachers' 'conversations' with children.

Providing opportunities for children to clarify meanings throughout is critical. They should be encouraged to keep mathematics diaries, and to make and to use mathematics dictionaries, problem-solving books, computer manuals, etc. - and to do so with confidence.

Language learning is a never-ending process. Children of all ages need support for meaning, through doing, observing and sharing. This is best achieved by providing an atmosphere conducive to hypothesising, doubting, qualifying, convincing, and other important functions of language. In such a climate, children initiate conversations and discussions with adults as well as with their classmates.

TWO ACTIVITIES TO ENCOURAGE DISCUSSION

The Round

The round is an activity that can be used to help children to talk in front of a large group and to increase their confidence that they will be listened to. The room needs to be organised so that the children are sitting in a loop of chairs and each can see everyone else's face. The teacher explains what is going to happen, in words something like this:

> "I am going to start a sentence and I would like
> each of you to think how you would finish it off.
> I will then ask each of you in turn to tell us what
> it was that you thought of. When anyone speaks,
> no-one else is to make a comment. I shall go round
> the loop asking each of you in turn; however
> you are allowed to say 'pass' and I will not mind.
> There will be quite a few sentences to complete, so
> I hope that you will not have to say 'pass' every
> time."

The teacher then starts a sentence such as, "What I liked about maths last year was ...", "What I have been doing this lesson is ...", or "I hope that I can get better at ...". As each child contributes, whether they pass or finish the sentence, the teacher acknowledges with "Thank you" or something similar. It is important that everyone, including the teacher, sticks to the rule about not commenting on any contribution.

Prompt Cards

The prompt card is a technique used by modern-language teachers to encourage children to talk to each other. Often this activity is in pairs, but it can be used with larger groups. It can be a useful confidence raiser if children have not been used to discussing their mathematics among themselves. The teacher needs to prepare some cards, with questions on them, which are related to the area that the children will be talking about. For example, if the discussion is to be about how they are going to write up an investigation, then a prompt could be "What are the results that we want to tell others about?" or "What did we do first?"

The idea of the cards, which the children can use by themselves, is to provide a means of support for the discussion, and so, to help them realise that they are not dependent on the teacher.

(These activities are each referred to several times in Part Two)

And Watching Your Language Details

There is no doubt that teachers can modify their language in order to encourage discussion.

Certain prompts are likely to stop children from continuing to talk; e.g. "Why don't you try using 3 counters?" "Shall we make a table?", or "Did you check it?".

Others will have the opposite effect; e.g. "What are you going to do now?", "Which was the hardest/easiest part?", or "How do you know it is correct?"

LEARNING LANGUAGES

The motivation for learning a language is the need to communicate. Children learning their first language spend about a year as active listeners tuning into the sounds of the language around them, learning to respond to different intonation patterns, gradually and tentatively trying out individual sounds, then groups of sounds, then their first words. Parents encourage their children to speak, but they do not (and cannot) force them to do so. Insisting on oral responses too early may hinder learning.

Bilingual children in this country are on the whole highly motivated to learn and to communicate in English as they recognise that they need it in order to survive and hopefully to thrive. It can be very worrying for an adult if a child who is in the process of learning a second language is silent; but, as with first language learners, it is unrealistic to expect response in the early stages, and we should not feel threatened by this. The child will almost certainly be learning a lot. Comprehension is always well ahead of production.

Second-language learners do not need to listen for as long before starting to speak; they have already learnt one language and they know a lot about how language works. They will not, for example, need to build up from sounds to words, but they do need to hear a lot of examples of the language being used so that they can start to construct their own model of the language.

In the classroom, teachers need to bear the above points in mind, and in addition to consider the following:

- the language used in the classroom should be the same as the language used elsewhere;

- language is best developed in the service of other learning;

- there is no textbook that can be used to teach a language;

- children make excellent teachers and helpers.

- Are the children repeating the language of the new concepts they are learning, by collaborating over their work, playing turn-taking games, re-telling stories, talking about and interpreting charts, etc?

- Do the books that you offer to the children provide clear visuals for discussion, images that they would wish to identify with, plenty of repeated/meaningful language?

- Are all the bilingual children contributing to the work of the class in a way that everybody can see and recognise?

This check-list describes ways in which bilingual learners can be supported in the classroom, yet at no point does it mention anything that is not already recognised as being part of good primary practice and beneficial to the whole class.

Many of these points will be covered if the children are working collaboratively and engaging in discussion together.

The following are further elaborations and examples for a check-list:

- In order for the children to build on what they already know, it is vital that, wherever possible, the teacher draws upon the cultural background of all the children when working through a topic, e.g. "When do you send cards to your friends?", "New Year", "Divali", "Christmas", "Eid", as well as reflecting these backgrounds in the materials that they choose to support the work.

- While it is important to recognise that bilingual learners will progress in English if they collaborate with English speakers, their grasp of concepts can be improved by being able to collaborate and discuss in their mother tongues.

The act of talking-while-working often means that the children will recall what was said, and will be able, with your help and encouragement, to use this for their writing. Back-up can be provided by taping the children, or by the well-tried method of keeping half an ear open and perhaps jotting down what you hear and think will be useful later.

Because peer group talk is so important for children's language development (first and second), it is essential that we organise our classrooms to take account of this. If we are lucky enough to eavesdrop on the extended discussions children have with each other, we cannot fail to be impressed by their linguistic range, (much of which we are not aware of in the more formal adult:child exchanges), and the extent to which they support each other.

There are many illuminating recordings of children helping each other: suggesting the appropriate word, modelling whole sentences for bilingual children, even differentiating the reading help they give to classmates according to the reading ability of the individual child. Children frequently understand what it is that blocks the learning of their peers, having only recently passed that way themselves. Collaborative small group learning is of great benefit to second language learners, but it needs to be structured and nurtured by the teacher. It is still not widely encouraged as a method of learning in school, and it may be viewed by children as 'cheating' in the context of other teaching approaches in the school. In addition, racism and sexism amongst pupils can make this a difficult learning model to establish. But where teachers have set out sensitively to establish collaborative learning, considerable learning and social achievements have been noted.

If you have bilingual learners in your class/group, or, indeed, children from differing cultural or linguistic backgrounds, have you checked the following points:

- Could the children feel alienated from a task because it assumes a cultural background which they do not have?

- Is it standard practice for the children in your class/group to work collaboratively and to discuss the task in hand?

- Do the bilingual children mix with the indigenous English speakers for this? And if not ....?

- Has plenty of clear visual support been provided?

- Have the children got aural support from at least one of the following:

    - Discussion with their peers
    - Tape recordings

PART ONE

IDEAS TO CONTEMPLATE

C. ACTUAL LESSONS.

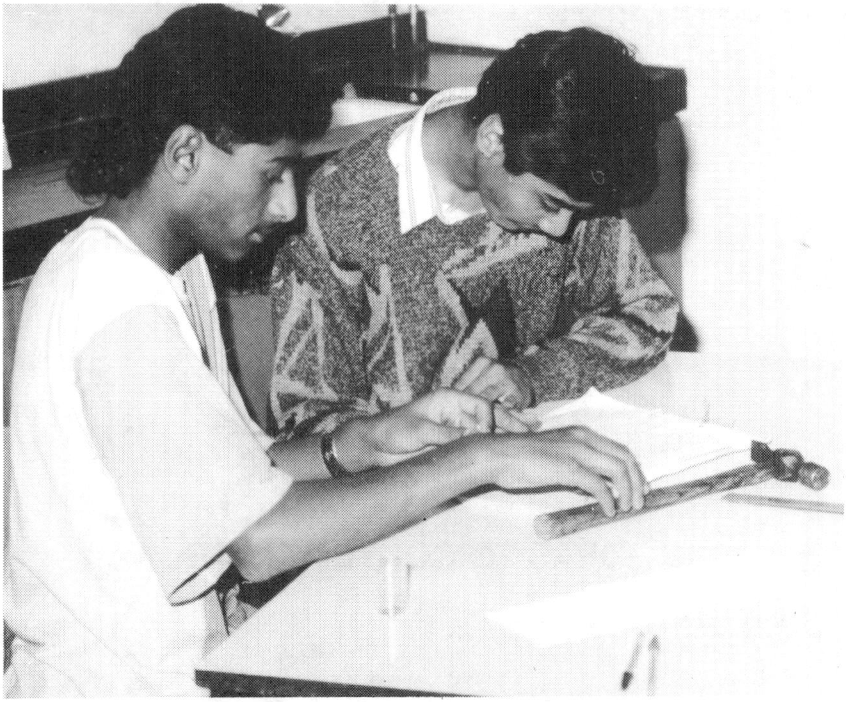

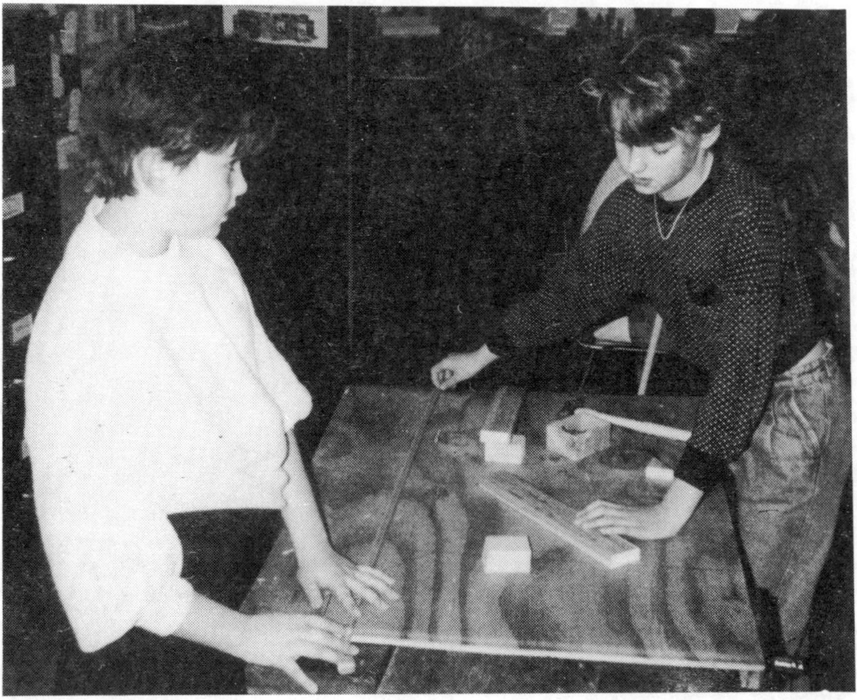

A SERIES OF LESSONS

The Class

The class consisted of twenty-five fourth year secondary children working towards a 50% coursework GCSE and hoping to get grades C to G. They had two 70-minute lessons a week as well as one homework which should last 45-60 minutes.

Lesson 1

I introduced the work by saying "I would like you to imagine a large field with a shed in it. The shed is 4 metres by 5 metres. A goat is tethered to one corner of the shed by a rope 3 metres long. What area of grass can the goat reach?"

After a few minutes of questioning, during which I repeated the information but did not give any further help on how to get started, some of the children got out rough paper and tried to draw a diagram. I went around the room helping individuals and making sure all understood what the problem was about. I made comments such as, "Can you explain your diagram?", and "What are you going to do now?"

After about 15 minutes, all of the class had arrived at a diagram which represented a top view of the situation. Some had drawn it to scale on squared paper and were counting squares, others were discussing the formula for the area of a circle.

I stopped the class and asked some of them to describe what they were doing. I had made sure that I would get some contributions, by giving praise wherever possible as I had gone round! Now I said, "I would like you to make up six questions about this situation which begin with, *What would happen if ...?*, for example, *What would happen if the rope were longer?* I am not going to ask you to answer all the questions you make up. Now you can all start with my example, so you have only five more to make up."

There was immediate discussion as children tried out ideas with their friends before writing them down. I went around, encouraging all of them to engage in the task and praising interesting questions. In particular, I warned a few of them that I would like them to share their ideas with the whole class. After about ten minutes all of them had at least four questions, so I asked for their ideas which I wrote on the board.

When the ideas started to dry up, I said, "I would like you to choose one or two of the questions that you have written down or have heard in the dicussion, and work on them."

The rest of the lesson was spent with children trying to develop their own particular lines of investigation.

Lesson 2

I started the lesson by asking the children to work in pairs. Each person had to describe to the other what questions they were working on, and how they had got started. I warned them that they should not be talking about answers but about strategies. After about five minutes, I asked the class to share some of the things as a whole. Then, back in their pairs, I asked each one to describe what they were hoping to do during that lesson, and finally this was shared with the whole class. By the time they re-started on the goat problem, most of them were clear what their strategies for the rest of the lesson were going to be.

The time left was spent working on the problem. All of the children were working in rough and I was encouraging them to write down as many of their ideas and findings as possible.

Lesson 3

As the children arrived, I encouraged them to start work on their particular questions straight away, and we spent more than half the lesson on this. I was encouraging general statements, and also developments of the original questions where the children seemed to feel they had answered these satisfactorily.

Towards the end of the lesson we had a whole class discussion based on general results, with some explanation from me highlighting the differences between particular results and general ones; and then we had contributions from children prompted by the conversations that I had had during the lesson.

At this discussion I raised the point of when the work could be completed and we agreed a deadline of three lessons hence.

Lesson 4

I started by reminding the group of the deadline we had agreed. Although they were experienced in writing up work for their coursework folio, we spent a little time as a whole class discussing points to remember, and also the alternatives open to them in terms of presentation: poster, booklet, etc.

About half the group started on their presentations straight away, but the others had some more work to do before they felt satisfied with their results. I spent the lesson trying to persuade some of them to continue with a little more research, and trying to persuade others to start their write-ups!

Lessons 5 and 6

These sessions were spent with the children completing their presentations. I brought to the room as many resources as possible and I helped with suggestions when the precise material wanted was not available. A lot of children had access to other resources outside of mathematics, but most wanted to complete the work in the room. I spent my time reading rough drafts and helping them to put into a written form some of the more difficult ideas that they had expressed orally.

As their deadline came closer, a small number of children came to me to re-negotiate their own personal targets; we tackled this on an individual basis, keeping the whole-group timetable as intended.

Towards the end of the last session many children had finished and I encouraged them to work on tasks from some of their other subjects.

We spent the final five minutes reflecting on the past three weeks work: what new skills had been developed, what felt good, and what was not so good.

A LESSON PLAN NETWORK

<u>Lessons from the Hundred Square</u>: a possible network of activities enabling students to use and apply mathematics.

- Fill in numbers 1 to 100 on a 10 by 10 grid.
- Choose a systematic method and describe it.
- Use Bengali numerals and translate it.
- Make up questions whose answers can fill the grid.
- Use mental arithmetic questions to fill the grid.

Each student has a hundred square, and two L-shaped pieces of card to make rectangular 'windows'.

Look at several 2 by 2 'windows' and describe any patterns you find.

- Make and test general statements: "In any 2 by 2 square...."
- Pool ideas in small groups or the whole class to help others get going
- Surprises

Some patterns involve +, −, ×, ÷. Others involve manipulating separate digits. Which are fruitful?

- Explain patterns to a friend
- Surprises

Algebra: some respond to a hint to use n, others will appreciate a fuller explanation, some will be content with a verbal explanation.

- If you can't find patterns, test someone else's ideas
- Surprises

What would you like to do next? You have the whole grid to work with.

- Use an isometric grid. What would you fill it with?
- Use 2 by 2 squares on a multiplication grid.
- Look at 3 by 3 and 4 by 4 windows
- Use rectangles
- Use L-shaped windows

General statements. Tests. Predictions. Surprises.

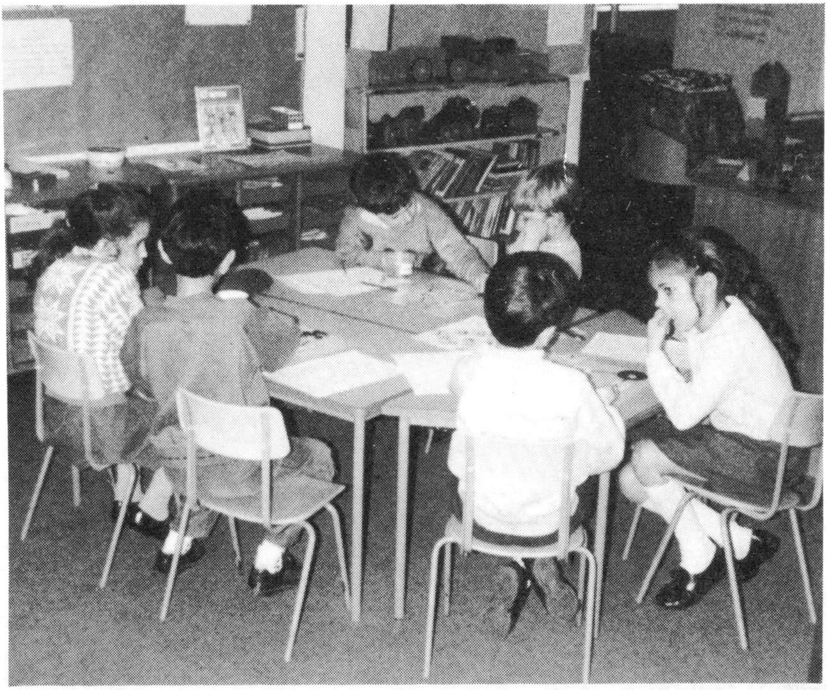

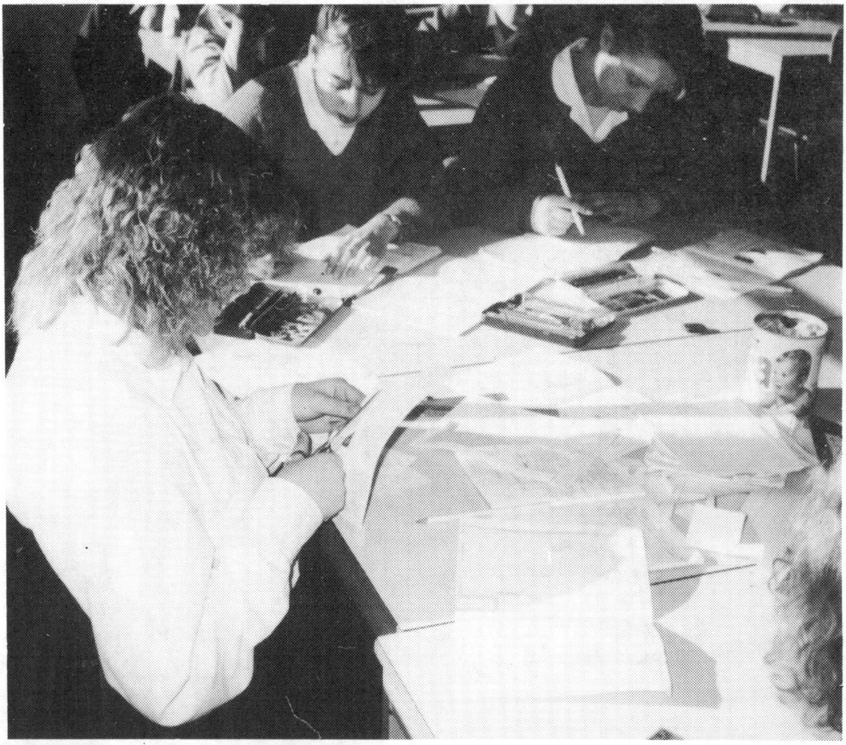

PART TWO

IDEAS TO USE

A. IDEAS ON USING MATHEMATICS

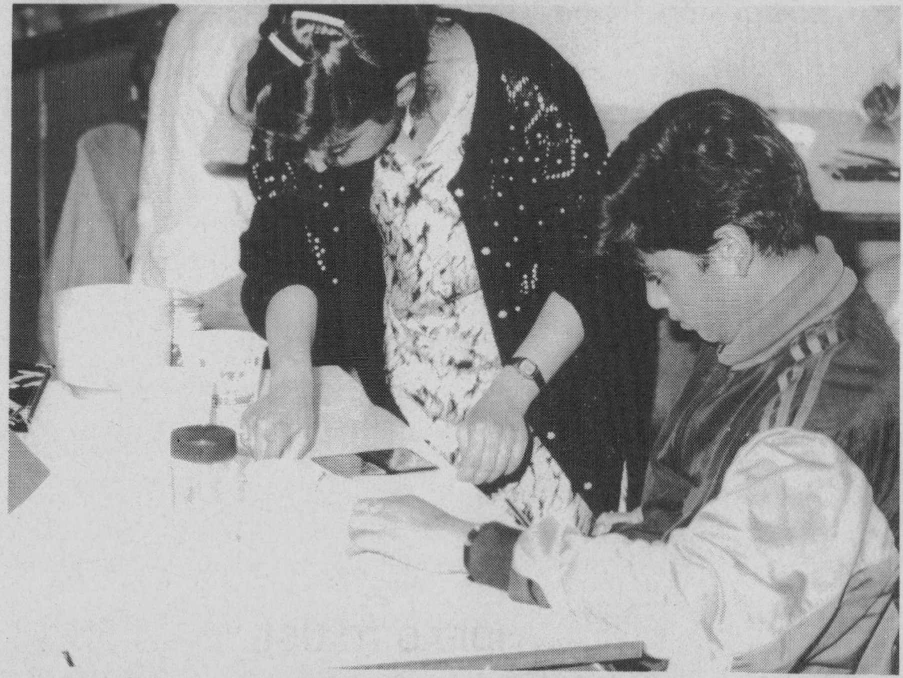

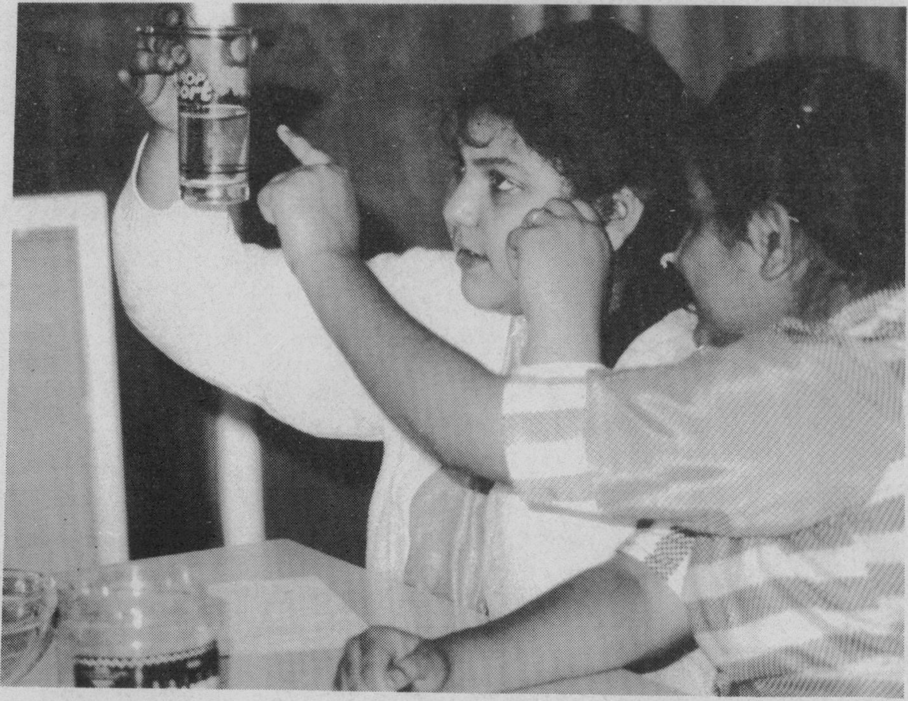

1 SELECTING MATERIALS AND MATHEMATICS APPROPRIATE FOR A PARTICULAR TASK

Strategies

- Starting with a restricted choice; e.g. colours of paint to use or types of graph paper, then gradually offering more and more.

- offering a wide variety of activities which stimulate the use of different media.

- treating all requests from children for materials or information seriously.

- thinking carefully about how organisation, both personal and whole school, allows for materials to be readily available.

- trying to decrease the number of occasions when children have to ask you for equipment.

- trying to praise children when they make independent choices. (This can be very difficult, especially when the teacher's choice would have been different.)

- having posters in the classroom or corridors reminding children of a range of approaches, processes or skills. These could be produced by the children themselves.

- trying to devise tasks which do not prescribe a particular method or end-point.

- encouraging children to see mathematics in particular situations and to feel it worthwhile to seek mathematical ways forward.

- trying not to say "no" to a request for materials, but being prepared to ask children why they have discarded alternatives. Not being afraid to include cost as a part of the decision-making process.

- being prepared for some losses, costing them into your budget and rehearsing with yourself and colleagues how you are going to handle them in class. Praising children who return equipment and being fairly low-key about small losses, is often more effective than heavy control.

- engaging children in discussions about which equipment, paper or method was the most useful and why.

- listening carefully to children explaining their own methods. This is mainly an individual or a small group exercise. You could use prompt cards to encourage discussion; e.g. What did you do first? Why?

- Encouraging conversation amongst children about how they plan to tackle a problem.

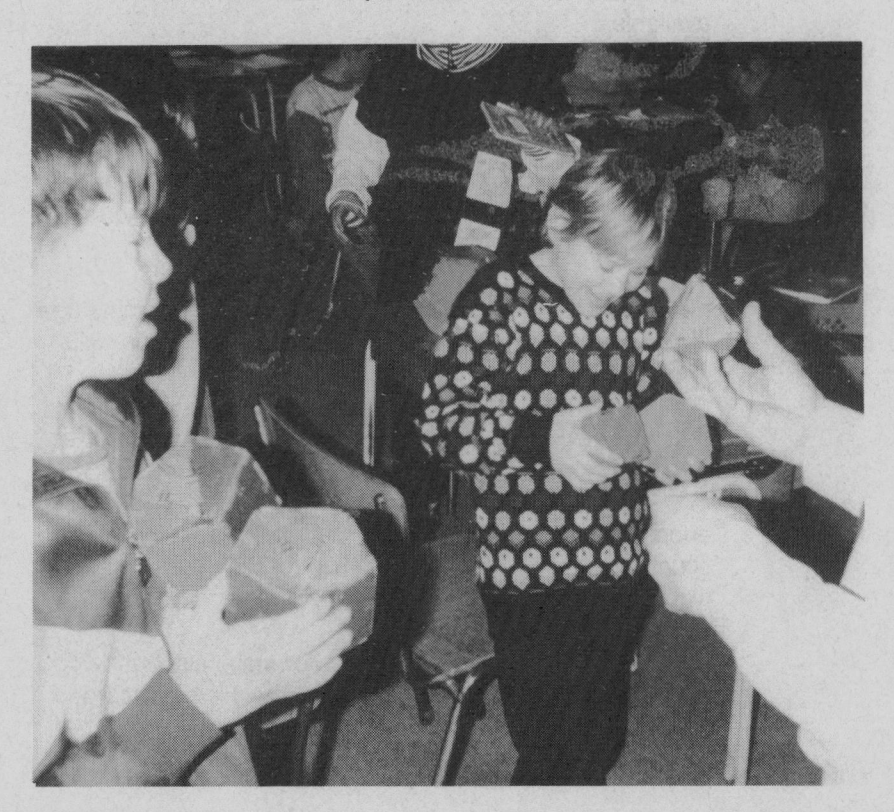

1. SELECTING MATERIALS AND MATHEMATICS APPROPRIATE FOR A PARTICULAR TASK

Activities

- "Find something heavier than 1 kg. Show me how you know it is heavier."

- having an estimation table in the classroom; e.g. "How many marbles in the jar?", "How heavy is the tin?" Asking children to explain how they made their estimate.

- designing a beaker to hold 300 ml of water.

- "How far will this ball roll?"

- "See how many different ways you can find to solve this equation ...."

2 PLANNING AND WORKING METHODICALLY

Strategies

- 'working methodically' may mean different things to different people. There needs to be respect for children's methods even though they may appear to be 'inefficient'.

- starting with conversations about a completed activity: "How did you do it?"; encouraging factual descriptions.

- sharing methods of attacking a particular bit of an activity; e.g. recording results or taking measurements.

- using reporting back to generate discussion about planning, possibly initially in a non-mathematical context; e.g. Planning for tomorrow: What do you need? - swimming kit, lunch, etc.

- encouraging children to talk to each other about how they plan to tackle a problem.

- encouraging children to record "What I am going to do next lesson" at the end of a session.

- before using the computer, encouraging children to plan the questions that need answering.

- presenting different methods for the same task. Asking children to compare, criticise and justify their choice of method; e.g. drawing a regular hexagon, or getting to school.

- asking "What can you change? What can you hold still?"

- asking "What do you think this is about?"

- involving children in negotiating ways forward.

- starting a task by brainstorming, or encouraging children to work a bit messily, and then drawing strands together to be coherent.

2. PLANNING AND WORKING METHODICALLY

Activities

- writing a set of instructions for a task; e.g. sharing out milk, getting up in the morning, or drawing a regular hexagon.

- writing LOGO code to draw a square, a triangle, a regular hexagon.

- inviting a 'guest' into the class. Planning what questions are going to be asked and by whom.

- devising a long-term task for the class to work on; e.g. keeping a weather record, or a record of what is eaten at lunchtimes. Allowing them to decide how to do it themselves.

- spending time on planning tasks, as an activity in itself, in situations where the tasks are not actually going to be carried out; e.g. building a house.

3   CHECKING FOR SUFFICIENT INFORMATION

Strategies

- trying to encourage some questions about information, such as:

    - "What happens if you ignore one piece of information?"

    - "What happens if you add or change a piece of information?"

    - "Will I really learn more from doing another one?"

    - "Do I need all this information?"

    - "Do I know enough to make a general statement?"

- asking if there is a more general problem lurking underneath the particular one.

- giving children problems in which the information is insufficient or ambiguous.

- giving children certain information and asking them what they could do, or find out, if they had more information.

3. CHECKING FOR SUFFICIENT INFORMATION

Activities

- using pictures of everyday objects cut into pieces, presenting them one at a time until the children can recognise the whole.

- playing attribute games with numbers; e.g. it is even, it is bigger than 10, it is the number of my house, etc.

- playing a Twenty Questions game with the whole class. Each child is allowed one question. They write down or quietly tell you the object when they think they know it. You record how many questions it took for each person, and encourage accuracy rather than pure guesses.

- trying a whole-group task where each member holds some information - some relevant, some redundant; sharing of information being done orally; e.g. The Zin Obelisk[1], or Murder Clues[2], or ...

- triangle clues: "you are describing a triangle to someone over the telephone, but are cut off after giving three pieces of information - were these enough?"

- looking for mathematics in newspaper stories, published statistics, atlases, etc.

- playing the function-machine game: your friend has to guess your rule.

1   From "50 Activities for Team Building" by Mike Woodcock. Gower Publishing Co.Ltd., Aldershot.

2   Murder Clues: Construct a murder situation with enough clues for each child to have at least one, but with more than you need for a solution. Include some irrelevant ones. The whole group has to communicate and to share information for a solution.

4 REVIEWING PROGRESS AT APPROPRIATE STAGES

Strategies

- using prompts, possibly on cards, with children; e.g.

    - "Tell me what you are doing"

    - "Tell someone else what you are doing"

    - "Are you doing what you set out to do?"

    - "Have you learnt anything from this?"

    - "What next?"

    - "Could you have done this last year?"

- encouraging children to note all thoughts; e.g. by drawing thought bubbles.

- encouraging awareness that mathematicians are 'lazy' and that they look for short cuts.

- suggesting to children that they make a mathematics diary entry at least once a week, answering the question "What am I doing at the moment and why?"

4. REVIEWING PROGRESS AT APPROPRIATE STAGES

Activities

- making a class chart showing which activities children are doing/learning. Asking them to fill it in.

- asking children to look back over a term's work; e.g. Where have they used 'timesing', bar charts, etc. (These could be on prompt cards).

- at the end of a piece of work, reviewing what mathematics was used, what strategies were useful, what didn't work, etc.

5 CHECKING THAT RESULTS ARE SENSIBLE

Strategies

- when children want to know if they are right, asking "What do you think?", or "Is it what you expected?", or "Is there any way that you could get a more accurate answer or measurement?", or "Can you think of another way to do it?".

- asking "Can you think of a way to check it?"

- resisting the temptation to mark everything right or wrong.

- encouraging an awareness that mathematics is elegant, and that we should try to avoid unnecessary complications.

- suggesting that children ask other children to check their results.

- asking children to "explain how you worked this out", both when you know that it is correct and when you know that it isn't.

- it is sometimes sensible to keep all questions in context, with numbers that make sense; but at other times it might encourage learning if one has unexpected answers; e.g. meaningless negative solutions to problems.

- encouraging children to familiarise themselves with all the different calculators that they use.

5. CHECKING THAT RESULTS ARE SENSIBLE

Activities

- putting out one pencil for each person in your group. How do we know that this has been done correctly?

- giving an answer and asking what the question was; e.g. "24", or "3 kilograms", or "A square".

- doing some money calculations on a calculator and asking "What does the display mean"; e.g. looking at £2.50 divided by 7)

- announcing that it took 7.319652 minutes to get to school this morning, and encouraging discussion.

- grains of rice on a chessboard: There is 1 grain in the first square, 2 in the second, 4 in the third, 8 in the fourth, and so on. How many grains of rice on the chessboard? (This leads to standard-form notation on the calculator).

- working out your age to the nearest second; discussing how to allow for the seconds/minutes that it takes to work out the answer.

6 USING TRIAL AND IMPROVEMENT METHODS

Strategies

- the important word to work on here is 'improvement'. This is what distinguishes a trial method from pure guesswork.

- encouraging children to think about their methods and to count the number of guesses needed. Discussing different methods and asking which needs the smallest number of guesses.

- drawing a flowchart for the method. Is there a loop, and if so, how long is it?

- introducing the idea of linear searches and of binary searches.

- trying not to act as judge yourself; encouraging the children to recognise errors themselves and to decide when they have a result that is accurate enough. "Are there things that you can change a bit to make the answer different?", "Is this too big or too small?", "Can you get closer?"

6. USING TRIAL AND IMPROVEMENT METHODS

Activities

- There are a number of computer programs which are based on this technique:

    - RHINO, ELEPHANT, and GUESS on SMILE[1] disks;
    - SUNFLOWER[2];
    - ERGO on SLIMWAM 1[3];
    - COUNTER on SLIMWAM 2[3];
    - The cake problem in L - A MATHEMAGICAL ADVENTURE[3].

- using the calculator to find two consecutive numbers which multiply to give 182

- before introducing the square-root key, using the calculator to find the square-roots of 25, 121, 169, ..., and also of 100, 200, 300, ...; doing a similar activity for cube-roots, fourth-roots, ...

- "Sort out these boxes ... Explain how you sorted them."

- "Find x such that $3x^2 - 3x - 1 = 0$ ". (Or with x^3)

1. SMILE "The First 31" and "The Next 17" are available from ILECC, John Ruskin Street, London SE5 OPQ.

2. SUNFLOWER is on "Teaching With a Micro : Maths 2" and on "The Language of Functions and Graphs"; obtainable from The Shell Centre for Mathematical Education, University of Nottingham, University Park, Nottingham NG7 2RD.

3. SLIMWAM 1, SLIMWAM 2, and L - A Mathemagical Adventure are published by ATM.

7 TRYING ALTERNATIVE STRATEGIES

Strategies

- valuing suggestions from children and not expecting too much detail.

- trying to act as chairperson for discussions about strategies; allowing children to recognise weaknesses themselves.

- using phrases like "What could we do now?" or "What do you think?"

- encouraging and accepting alternatives in any situation - not only when you are expecting them.

- encouraging mental methods as a starting point.

- encouraging pictures and diagrams of a situation.

- other strategies include group-work, pooling ideas as a class, children-generated lists of what was useful in a task.

- using display space in the classroom, (e.g. a work-in-progress board), where children can pin work at the end of a session instead of putting it in their bags to take away.

- in the initial stages, trying to accept work uncritically; expect to feel a little insecure!

- suggesting that children go back to the starting point, or to their last decision, and re-think.

- being aware that the way a problem is presented can restrict the perception of children; e.g. a geometrical problem may be solvable algebraically.

7. TRYING ALTERNATIVE STRATEGIES

Activities

- making up some number sentences with the answer "4".

- "How many presents would you receive on the 12th day of Christmas?"

- "How high is the building you are in?"

- continuing. the. sequence. below. .for. two. more. terms: generalising the idea ...

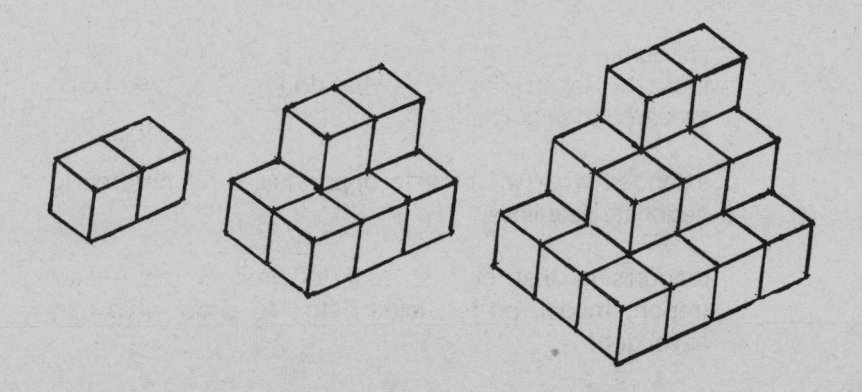

8 COMPLETING A TASK

Strategies

- helping children to recognise that they can be in control of what they learn.

- giving 'open-ended' tasks in order to encourage open-mindedness.

- encouraging children to decide for themselves when a piece of work is finished, but being prepared to suggest ways forward yourself.

- sharing with children the aim of the task, so as to give it a sense of worth beyond just pleasing the teacher.

- accepting that too rigid an insistence on finishing can dampen enthusiasm.

- encouraging children to say "Do I have to do all of these?". Responding with "What do you feel?" and being prepared to negotiate.

- extended work will provide opportunities for children to negotiate deadlines.

- recognising that children like to produce an artifact (report, model, poster, folder, etc.) to show what they have 'done'.

- encouraging children to complete their work by listing possible extension activities.

8. COMPLETING A TASK

Activities

- "How many routes are there along the edges of a cube between two particular vertices?"

- planning a class party to happen on a certain day - and making sure that the event then happens!

- producing a report or poster about the important points in a piece of extended work.

- convincing someone else that the angle in a semi-circle is equal to 90 degrees.

- concluding a task by listing extension suggestions.

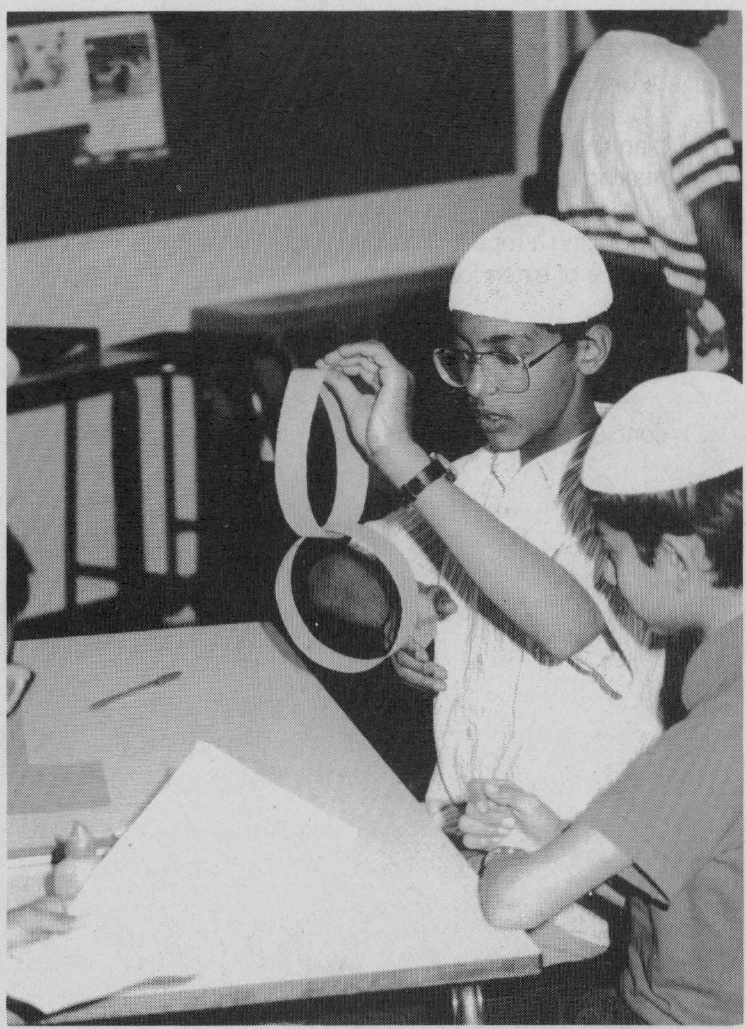

9 PRESENTING ALTERNATIVE SOLUTIONS

Strategies

- discussing the nature of accuracy, in order to help children recognise when different levels of accuracy are needed; e.g. in measuring tasks.

- making explicit the mathematics involved in tasks which are not obviously connected.

- trying to value different interpretations of a problem.

- asking questions like "What rules did you make for yourself?" or "In what situations is this true?" or "Can you justify this in terms of the original problem?"

- presenting some real-life tasks, perhaps in which there are outside constraints; e.g the best route for delivery vans making multiple deliveries, or the number of lifts needed in a building; perhaps linking with Industry.

- presenting open-minded tasks in which only partial solutions are possible.

- presenting tasks which need some definition or some assumptions by the child; e.g. "How many squares can you see on a chessboard?"

9. PRESENTING ALTERNATIVE SOLUTIONS

Activities

- choosing tasks in which aesthetic, moral or practical judgements have to be made; e.g. designing a bedroom, spending £1 for lunch, or planning a school fete.

- "Cherwell's poppadoms are in packs of 4, Joy of Delhi poppadoms are in packs of 8. Both packs weigh the same. Compare the sizes of the poppadoms."

- devising a fair way of giving each child a turn to take responsibility for something.

- finding the 'best' place for 7 people coming from 7 different parts of the country/town/school to meet.

- making explicit the idea that there may be more than one way of getting 'the right answer'.

PART TWO

IDEAS TO USE

B. IDEAS ON COMMUNICATING MATHEMATICS

MAKING SENSE OF A TASK

Strategies

- giving tasks orally: this provokes children to decide what they need in order to make sense of them, and it encourages them to ask questions.

- making sure that the children have heard and understood the words before offering to help with the task.

- trying to resist the temptation to move from explaining the task onto making decisions about how it should be tackled.

- trying not to give too much detailed help initially; instead waiting for the children to ask.

- using others to describe (and devise?) a task; e.g. another teacher, the caretaker, a child, a parent, a worker from an industrial company.

- asking children to make up a problem or to bring one in from home.

- using problems set in magazines, books, or other media.

- using tasks that involve sequencing; e.g. making a cup of tea, or getting dressed.

- thinking about, and rehearsing with colleagues, how you are going to respond to children's insecurities expressed by "Is this right?", "I'm stuck", or "What do I do now?"

- encouraging children to talk about how they got started on a task. This could be part of a class review discussion.

- encouraging children to "put it in your own words", or to "try a few numbers and see what happens", or to "draw a diagram".

10. MAKING SENSE OF A TASK

Activities

- using tasks in which it is easy to understand what is needed, but which are not so easy to complete; e.g. design a board game or a children's puzzle, or "How many squares are there on a chessboard?"

- using the Hundred Square poster *(published by ATM)* and other posters, in order to stimulate discussion and to raise questions.

- "There is a streetlamp. A child one metre tall walks by. What happens to the top of their shadow?"

- "Plan your ideal school timetable."

- "Rearrange the tables in your classroom."

11   INTERPRETING MATHEMATICAL INFORMATION

Strategies

- this is related to checking for sufficient information.

- trying to use real information as far as possible; e.g. real timetables, catalogues, newspapers, magazines, or children's games - even if you restrict access to only part of them.

- the difficulty of interpretation often lies in the amount of distracting information in, for example, a table or a graph. Try starting with as little as possible and yet still be realistic.

- using sources of information from other curriculum areas; e.g. humanities or sciences. There is a real opportunity for collaboration here.

- using computer software which involves interpreting relevant information; e.g. Global Statistics[1], or Domesday[2].

- encouraging children to see that there is mathematical information everywhere around them and that it is not just in numerical form.

---

1. *Global Statistics by Brian Hudson is available from the Centre for Global Education, University of York.*

2. *Domesday - the interactive video disc.*

11 INTERPRETING MATHEMATICAL INFORMATION

Activities

- making up a story to go with a number sentence.

- sorting out patterned wallpapers.

- looking for and describing tessellations in your home, school, journey in-between.

- "Red Alert! This building is the centre of a toxic chemical leak. You have got to get as far away as possible in one hour."

- "Make up five mathematical questions about this room."

- using a spreadsheet to solve the problem of producing a box of maximum volume from a given size of card.

- using a graph plotter, e.g. Omnigraph[1], or FGP[2], to investigate the coefficients of a polynomial.

- children can read statistics from a publication, e.g. a daily newspaper, *Internationalist* or *Current Trends* (published by HMSO), and choose an area to talk about. Data gathered from the classified columns of local newspapers can provide interesting studies (for example) on the car most often advertised.

---

1. *Omnigraph by P R Brayne is available from Software Production Associates, P.O. 59, Leamington Spa, CV31 3QA*

2. *FGP is on SLIMWAM 1, published by ATM*

12 TALKING ABOUT WORK IN PROGRESS AND ASKING QUESTIONS

Strategies

- in class discussions, encouraging general descriptions of what the children are (or have been) doing, rather than descriptions of specifics; e.g. when they are drawing up a table, or trying to find a pattern, or writing an introduction.

- using a round with starters like "What I have done today is ..."

- using prompt cards to encourage groups to discuss their work; e.g. "How do you know that is right?", "What have you done so far?" or "Why did you do that?"

- when talking to individuals, starting with general prompts such as "What have you been doing?", "Why?", "Why not?", or even "What if ...?" Then focussing in to more detail. Encouraging the children to take responsibility for raising the next question.

- trying not to automatically answer questions that children ask. Trusting them to have enough knowledge about a situation for you to help them to discover something for themselves. (However, there are times when a quick answer is more appropriate).

- having a work-in-progress board where children pin their work instead of putting it away.

- listening uncritically to children's discussions.

- trying not to interrupt discussions: the children may not need you.

- a tape recorder left on the table can help to focus children's minds.

12. TALKING ABOUT WORK IN PROGRESS AND ASKING QUESTIONS

Activities

- "Sort out these teddy-bears/boxes/shapes/extracts from newspapers[1]. Explain how you did it."

- "Write some questions to display with your bar chart."

- designing some prompt cards for finding the area of a triangle.

- inviting a guest (another teacher, an industrial contact, an older child, a parent) into the classroom to talk with individual children about their work.

- designing a display of current work for a Parents' Evening.

---

1. *If you wish to use a newspaper in the classroom the following below will be able to help with ideas:*
   The Northcliffe Newspapers in Education Project, Tower Lodge, Sandown Park, Tunbridge Wells TN2 4RH (Tel.0892 512321)
   The NIE Manager, The Newspaper Society, Bloomsbury House, Bloomsbury Sq.,74-77 Gt.Russell Street, London, WC1B 3DA.(Tel.01 363 7014)

13. EXPLAINING AND RECORDING WORK SYSTEMATICALLY

Strategies

- trying to value rough working and to encourage children to keep it. Can it be accepted within a part of your assessment schemes?

- encouraging children to record work in tables when this is sensible. Allowing them to design their own and then to reflect on the design.

- children could study systems in use elsewhere; e.g. in a dictionary, telephone directory, or library.

- devising tasks that help children to recognise variables; e.g. when watching television or drawing a square.

- being systematic is about controlling variables. There are computer programs which allow this; e.g. LOGO or TILEKIT[1]

- encouraging all approaches - efficiency can be considered later.

- using a camera, a tape-recorder, or video tape to record work.

---

1. *TILEKIT is on SLIMWAM 2 published by ATM*

13. EXPLAINING AND RECORDING WORK SYSTEMATICALLY

Activities

- finding rectangles with an area of 24 squares.

- making up and recording a rota for taking the gerbil home.

- "If someone bought house-numbers 1, 2 and 4, what could the number of their house be? If these and one other house-number were bought together to do for two next-door neighbours, what could the numbers of these two houses be?"

- on squared paper, drawing a rectangle, and then drawing a diagonal in it.

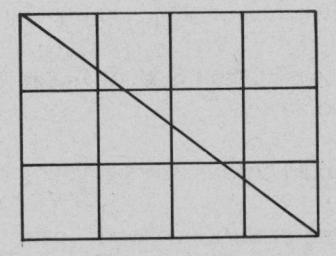

  Doing the same for 5 more different rectangles. Sharing them with a friend, sorting them out and deciding which type to investigate. *(See Points of Departure 1, published by ATM)*

- "Explore $y = ax + b$", recording your findings. What do *a* and *b* do?"

- "Some calculators have an AND button. What does it do?"

14. PRESENTING RESULTS IN AN INTELLIGIBLE WAY TO OTHERS

Strategies

- encouraging children to consider a wide variety of presentations; e.g. models, OHP, blackboard, poster, booklet, or even a lesson.

- making time for discussion of presentations in a critical way; e.g. of a TV programme or a magazine article. What was good? What could be improved?

- encouraging group presentations and the making use of other peoples' skills.

- children should ask themselves "Is this necessary? Does it help me, or others, to understand?"

- a presentation of work should have a beginning, a middle and an end.

- should children work in their mother tongue?

- word-processing work can encourage children with writing difficulties.

- 'intelligible' does not necessarily mean 'neat decorative work'.

- 'intelligible' may not mean using classical mathematical language.

- but the importance of using conventional mathematical terms must be recognised.

- the production of work for display can involve mathematics which is only implicit in the finished product.

- talking to English teachers about how to encourage children to present their written reports.

14. PRESENTING RESULTS IN AN INTELLIGIBLE WAY TO OTHERS

Activities

- children can produce a poster listing many types of mathematical communication: graphs, algebra, number, etc., for future reference.

- looking at mathematics from other children, other times, other cultures in order to learn about other ways of communicating mathematics.

- listening to a teacher reading out a child's record of work which contains elements of good communication. This should not necessarily be the 'best' possible piece of work but something that children can aspire to.

- reproducing a pattern based on circles, but drawn to a different size: e.g. twice the area (How should the radius be changed?)

- designing a display of current work for a Parents' Evening.

- getting older children to test out theories written by younger children and to write them a reply.

- "Can you work with your English teacher to produce a report?"

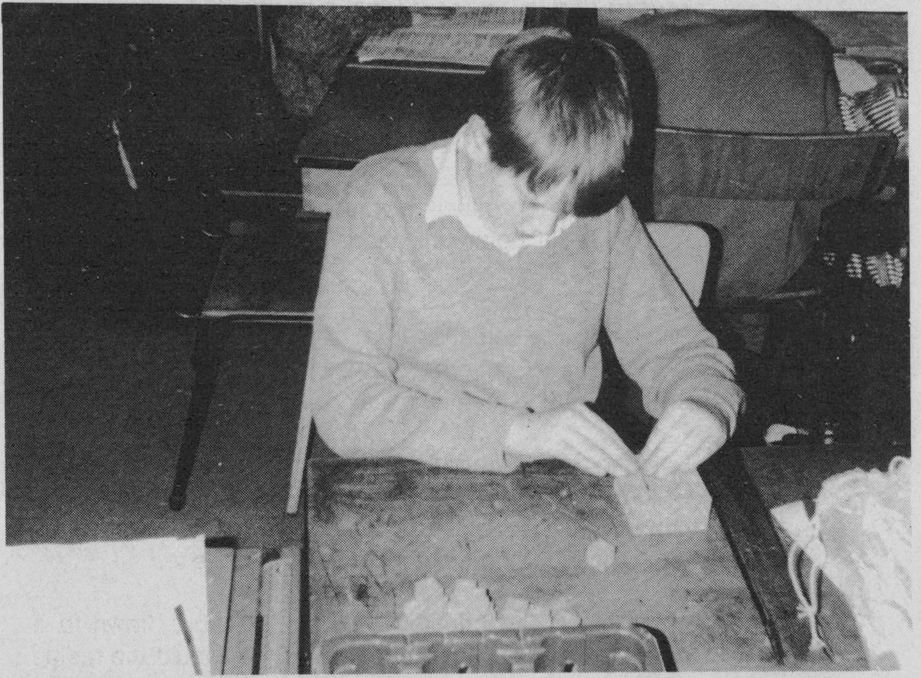

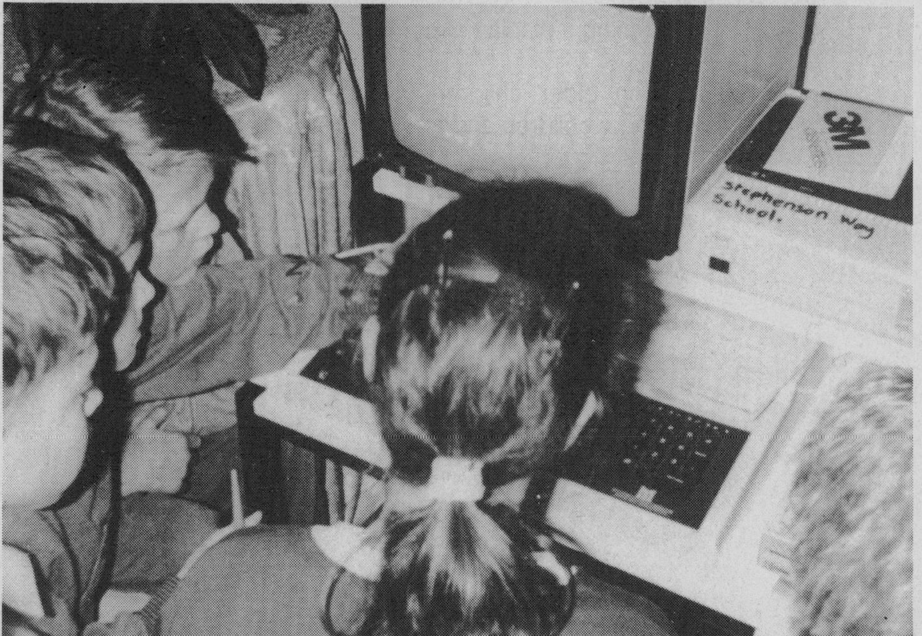
3M
Stephenson Way
School.

PART TWO

IDEAS TO USE

C. IDEAS OF DEVELOPING ARGUMENT AND PROOF

15 ASKING THE QUESTION "WHAT IF ...?"

Strategies

- having times when the purpose is to think up "What if ...?" questions, rather than to answer them.

- being careful to point out that you will not ask children to look at *all* of the questions that they come up with. This will help overcome any tendency they have to only ask questions to which they already know the answer.

- valuing all questions; writing them on the board in class situations, or encouraging children to write them down individually if they are working in small groups.

- encouraging mathematical curiosity in the classroom, by using posters, puzzles, etc.

- the examples in the Attainment Targets are not very helpful; better ones might be:

  AT1 Level 2 - having found the capacity of a cylinder, make up questions like "What would happen if it were taller, shorter, fatter, ...", and then answer some of them.

  AT9 Level 2 - having completed a survey of favourite music in their own class, decide who else they could ask and what difference it might make to the results.

- defining the process 'developing a problem' for the children so that you can help them to increase their effectiveness as investigators.

- at the end of an investigation or a game, asking "What if the rules were different?" Making this a standard part of writing up investigational work.

- "What if ...?" questions are not restricted to developing ideas of argument and proof, but are also useful when a child seems 'stuck' and also in changing 'closed' tasks to 'open' ones.

15. ASKING THE QUESTION "WHAT IF ...?"

Activities

- "Painted Cube" (*from ATM's Points of Departure 1*) ... What if it were a cuboid ...?

- ask the class to write down six questions about the problem which they are working on, each question starting "What would happen if ...". This is *always* possible, even with the most closed-seeming problem. If necessary, giving one example to get them started.

- Billiards/Rebounds/Snooker (it is known by a variety of names) can allow children to ask "What if the snooker table is a different rectangle? ... has non-integral sides? ...is a parallelogram? ...has the pockets in different places?"

- What if we feed the gerbil every two days, how much should we give it?

- When starting on a problem, writing down lots of questions without stopping to think, and then choosing one or two to work on.

- How far away from (0,0) is a given point? What if the point travels along a straight line? ...a circle? ...an ellipse?

- "What if not ...?"

MAKING AND TESTING PREDICTIONS

Strategies

- children should be encouraged to see that making and testing predictions is a highly significant activity which forms the basis for all scientific knowledge.

- the use of investigative situations will offer many occasions for predictions; however, be aware that a predicting activity may turn into merely guessing, without any checking and improving.

- encouraging children to voice predictions out loud. Talking about the importance of them saying when they are not sure.

- being careful to value all suggestions equally. Children need support in making realistic predictions. Initially they will have little experience on which to base them.

- young children will have problems making distinctions between fact and fiction. Asking them to justify their predictions will be threatening, and it needs to be handled carefully.

- helping children to recognise that they are doing good mathematics by commenting "That's a nice prediction" whenever possible, irrespective of whether the prediction happens to be true or false.

- encouraging involvement, by asking children to write down their predictions and *then* working to see if they were right.

16. MAKING AND TESTING PREDICTIONS

Activities

"Each point on a circle is joined to each other point. The resulting regions are counted

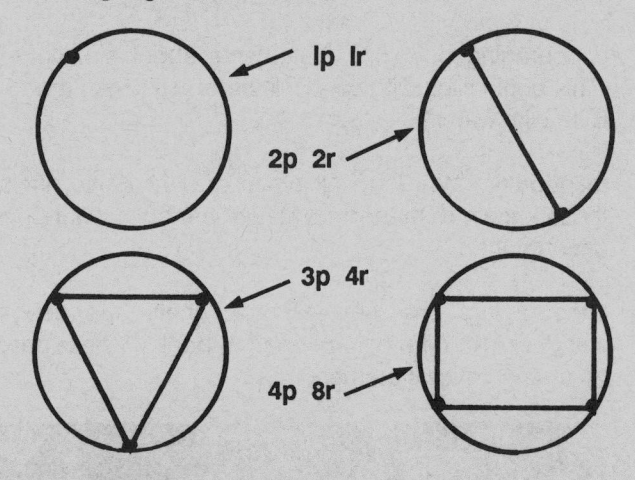

1p 1r
2p 2r
3p 4r
4p 8r

"How many r for 5, 6, 7, ... points?"

- "What will happen if I take this brick from the balance?"

- "If we arrange cubes to make a step pattern, how many will there be in each layer?"

- "If everyone in the room shakes hands with each other person once, how many handshakes will there be?"

- Using population figures from 1970 to 1985 to predict the population in 1992.

- Investigating the effect on a matrix transformation of interchanging two elements in its matrix.

17 MAKING AND TESTING STATEMENTS

Strategies

- resisting the temptation to act as a judge of statements, (especially this one).

- encouraging particular statements about situations, e.g. "This bottle filled with sand will be heavier than it would be filled with water".

- responding "Why?" to all types of statements, including those known to be correct as well as those where there is uncertainty.

- encouraging children to cross-check corresponding statements in different information books; (these need not be mathematics statements).

17. MAKING AND TESTING STATEMENTS

Activities

- asking children to classify a bank of statements as:

    - 'fact' or 'fiction'

    - 'true', 'false', or 'possible'

    - 'always true', 'sometimes true', or 'never true'

    - 'propositions' (i.e. statements that are either true or false:. e.g. A square has four sides), or 'non-propositions' (e.g. Red is a nice colour)

    - 'testable' or 'non-testable'

    (See activity card games in "Whatever Next", published by ATM)

- asking children to look through their own workbooks and to collect examples of different types of statement that they have used in the past.

- "Choose pieces of information that you could give about a triangle. Which three pieces would be enough to describe the triangle completely?"

- "The playground is 30 big paces wide. Is this correct"

- " 'If $f(x)$ is negative when $x=0$, and $f(x)$ is positive when $x=2$, then there is a number in between which makes $f(x)=0$' Can you find counter-examples?"

18 GENERALISING, MAKING AND TESTING HYPOTHESES

Strategies

- helping children to recognise when they are generalising, by pointing out general statements when they occur, or by pointing out how to change a specific statement into a general one; e.g. "This bottle is heavier when filled with sand; all bottles are heavier when filled with sand."

- identifying the link between hypotheses and "What if ...? questions.

- in statistical work, encouraging children to make testable hypothetical statements and to then set about finding the evidence; e.g. "All people with big feet like mathematics", or "Young people are becoming more influenced by the media".

- encouraging children to describe number patterns; e.g. in a table of results, or in the multiplication table.

- generalisations like "All multiples of 5 end in 0 or 5" are accessible to most children. Be careful not to make the generalisation for them!

- encouraging the use of algebra to describe a general statement wherever possible.

- asking questions which encourage conjecture, like "Why do your results go up by 7 each time?", or "What would happen if we could repeat this process for ever?"

18. GENERALISING, MAKING AND TESTING HYPOTHESES

Activities

- asking questions which encourage generalisations, like "How many triangles in the 100th pattern?", "Is 20796842165 odd?", "is 33333337 a prime number?" "How many intersections are possible with 100 straight lines?"

- encouraging pupils to test hypotheses by asking questions like "Are girls better at maths than boys? How could you find out?"

- "Painted Cube" *(from ATM's Points of Departure 1.)*

- arithmograms:

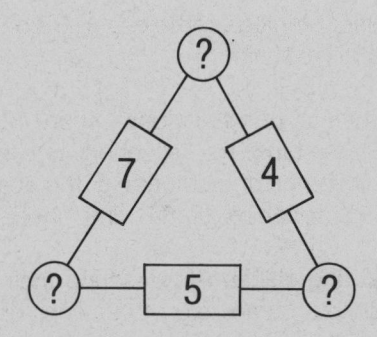

"What numbers can be put in the circles and added in pairs to give the numbers in the boxes? ... Invent more".

- when any question or problem is resolved, trying it again with a changed parameter; e.g. using hexagons instead of triangles, or using 6, 7, 8, ... instead of 5)

19. FOLLOWING ARGUMENTS AND REASONING, AND CHECKING FOR VALIDITY

Strategies

- asking children to explain 'Why' as much as possible.

- using prompt cards to encourage children to ask each other the same question.

- using phrases like "How do you know that?", even when the child's statement is known to be correct.

- trying to distinguish statements that are based on fact from those based on judgements. Value both, but expect different forms of reasoning.

- encouraging children to question each other's statements and reasoning.

- setting up situations where the children can easily understand the situation but where there may be a difference of opinion; e.g. the best way of spending £1, or the probability of throwing 2 heads, 3 heads, 4 ...

- asking children to describe their strategies for winning a simple game; e.g. 4-in-a-line, or Noughts and Crosses. Encouraging them to argue the case for their strategy.

- asking the question "How do you know you have got them all?" in counting situations; e.g. with "How many children in the class?", or "How many squares on a chessboard?"

- encouraging an atmosphere in class explanations in which children feel happy saying "Hold on a minute". Maybe using an agreed signal when they are not sure. Usually children only put up their hand when they *are* sure.

- valuing 'trial and improvement' as a powerful strategy.

19. FOLLOWING ARGUMENTS AND REASONING, AND CHECKING FOR VALIDITY

Activities

- "Tell a friend what your favourite food/subject/book/number is, and explain why"

- using DICECOIN[1] and asking the children to explain to each other the graphs that they get.

- using a local map, give instructions to get to places; can the children find where they are?

- playing Guess My Number; e.g. "My number is less than 10 but bigger than 7, and it is an even number."

- sharing out counters among a group of children: "What will happen if the number of children or the number of counters changes? - Give or ask for an argument for what numbers would enable you to share them out fairly."

- "if I multiply all the lengths of a shape by 2, then the area should be multiplied by 4 ...".

1. *DICECOIN is on SLIMWAM 1, published by ATM*

20 CONJECTURING, DEFINING, PROVING AND DISPROVING

Strategies

- Proof is a hard process. Make sure that the situations in which you ask children for a proof are easy to understand.

- the computer software QED, (*published by ATM*) is designed to encourage proof.

- offering opportunities for children to define their own variables whenever possible; e.g. when developing problems in words into algebra.

- asking children to say what they think will happen and to commit themselves by writing down their suggestions.

- getting children to look for causes and for explanations by constantly asking "Why?"

- using the word 'conjecture' yourself in class discussion.

- encouraging children to define problems clearly by being purposefully vague yourself; e.g. "How many squares on a chessboard?"

- encouraging children to define terms clearly, by being purposefully awkward yourself; e.g. by asking a pupil to define a rectangle and then drawing a non-rectangle that fits their description.

- choosing activities that start with a definition, after suitable negotiations as to its meaning.

- choosing activities that gradually refine a definition.

- accepting that in nature proof is relative rather than absolute.

- encouraging pupils to explain to each other, to you, and to themselves.

20. CONJECTURING, DEFINING, PROVING AND DISPROVING

Activities

- discussing logical fallacies such as "All dogs have four legs, a cat has four legs, therefore a cat is a dog".

- asking pupils to design a classification tree; e.g.:

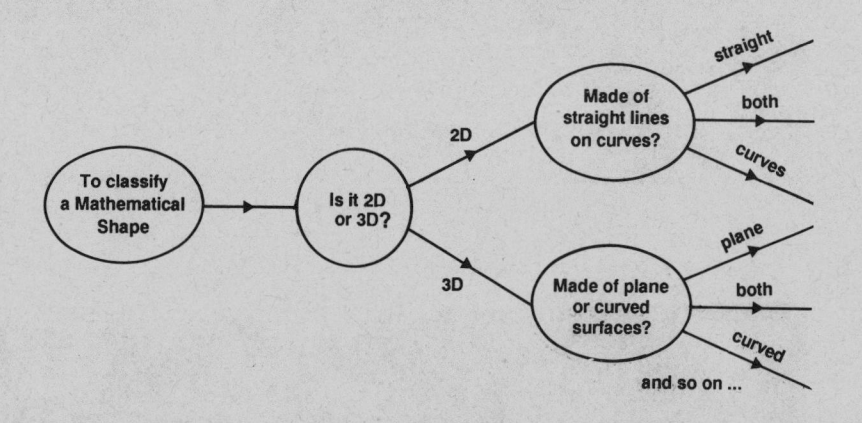

```
To classify a Mathematical Shape → Is it 2D or 3D?
    2D → Made of straight lines or curves?
        → straight
        → both
        → curves
    3D → Made of plane or curved surfaces?
        → plane
        → both
        → curved
and so on ...
```

- triangle clues: How few/many pieces of information do we need to define a particular triangle?

- "When is the angle in a segment of a circle equal to 90 degrees? Why?"

- "Everyone is going to find a place to stand that is 2 metres from me. What shape will we make? ...Why?"

- "There are 5 balls in this bag: 2 white and 3 green. Which colour might I pull out?"

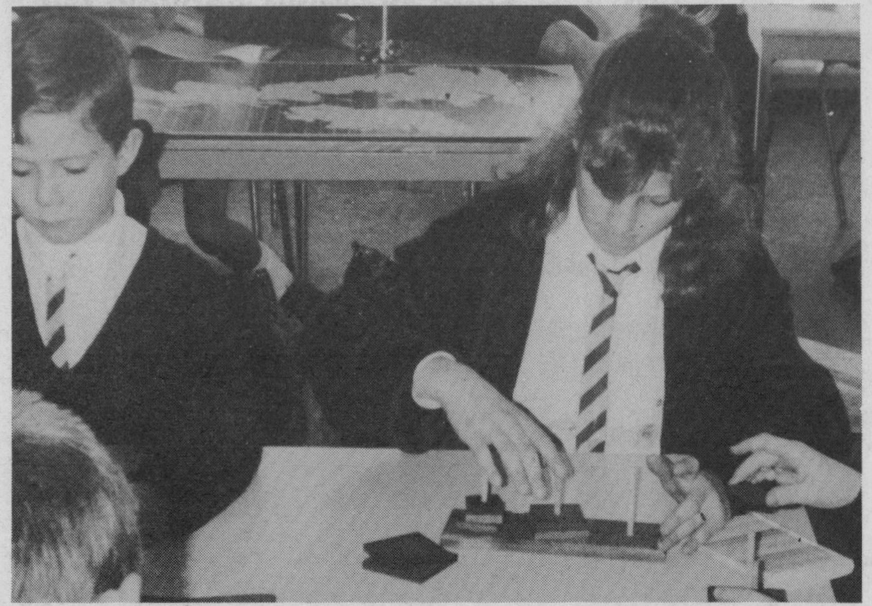

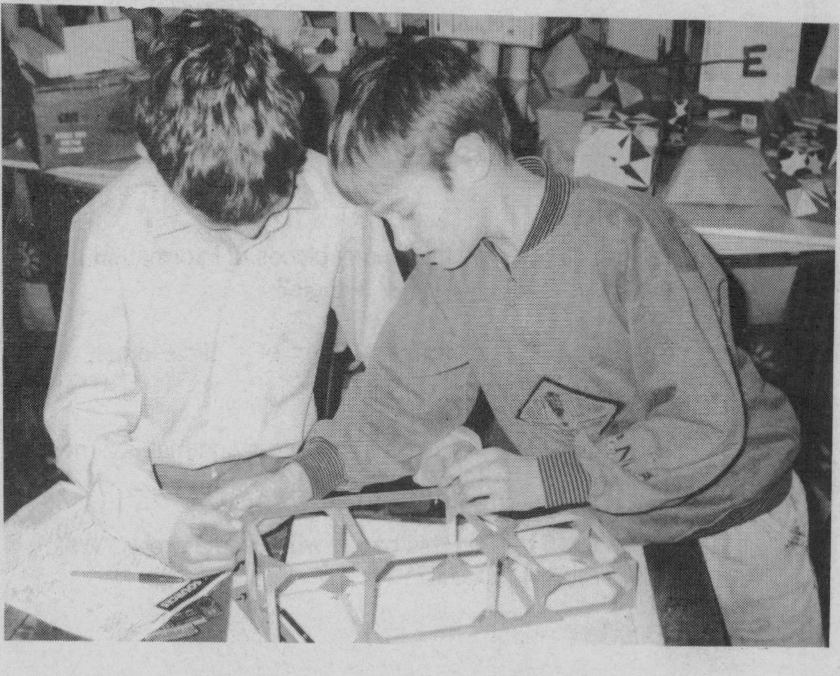
E

NOTES ON THE CHARTS

The charts accompanying this booklet show how the Statutory Orders attempt to map nineteen identified strands in Using and Applying Mathematics through the ten attainment levels.

The entries are taken from the programmes of study, excepting for the words in [square brackets], which help to clarify, and which are taken from the attainment targets.

When the Orders repeat a phrase exactly, the later statements of the phrase are omitted.

Observe the strong diagonal element within each of the seven groups of related strands.

What goes in the empty cells?

The apparent anomalies could be explained as follows:

Designing

L6 - L7/9   'a task' becomes specifically 'a mathematical task'. (c.f. how, under Selecting Materials/ Mathematics L1/2 - L3, 'a practical task' becomes more generally 'a task').

L7 - L9   'devising' becomes 'designing'. This is reasonable if 'devising' is having an idea, and 'designing' is mapping out the idea.

Working Methodically

L5 - L7   Working 'within an agreed structure' is more demanding than working as one wishes.

Communicating

"Oral, written or verbal"

The considerable repetition of this phrase is due to its entry under several strands, (Interpreting, Recording, Presenting), corresponding to different aspects of use.

Note the distinction between 'form' and 'forms'. In particular, in Presenting Work L4 - L6, the move is presumably from using the one form that 'falls out of the activity' to using any of the forms irrespective of how obvious their use might be. On this interpretation, 'or' at L4 is the exclusive 'or', but 'or' at L6 is the inclusive 'and/or'.

Recording

L3 - L4

If we view 'recording' as holding, rather than as presenting work, i.e. as what one does for oneself rather than for others, then initially a child will need the help that is given by being systematic; later the child can be more haphazard (as we are!).

Defining

L6 - L8

This makes sense if at L8 we interpret 'defining' as defining with full precision.

Reasoning

L6 - L8

This makes sense if at L8 we interpret 'reasoning' as reasoning with full precision.

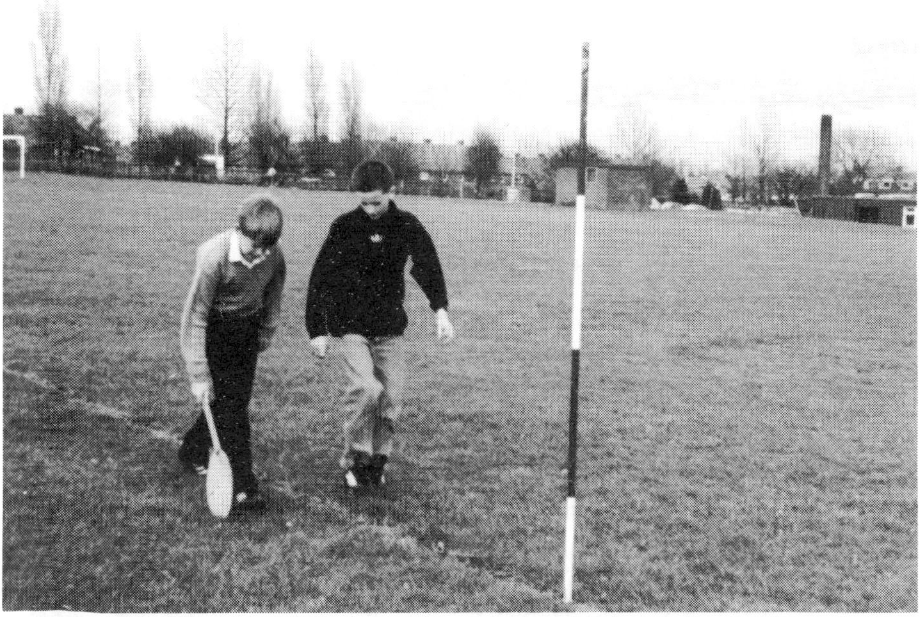

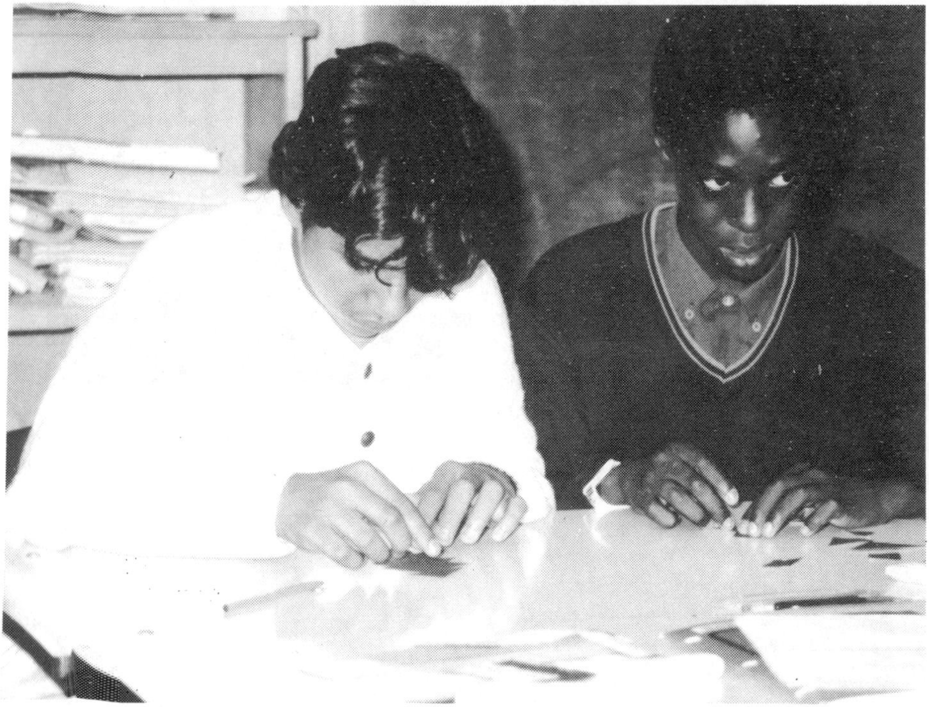

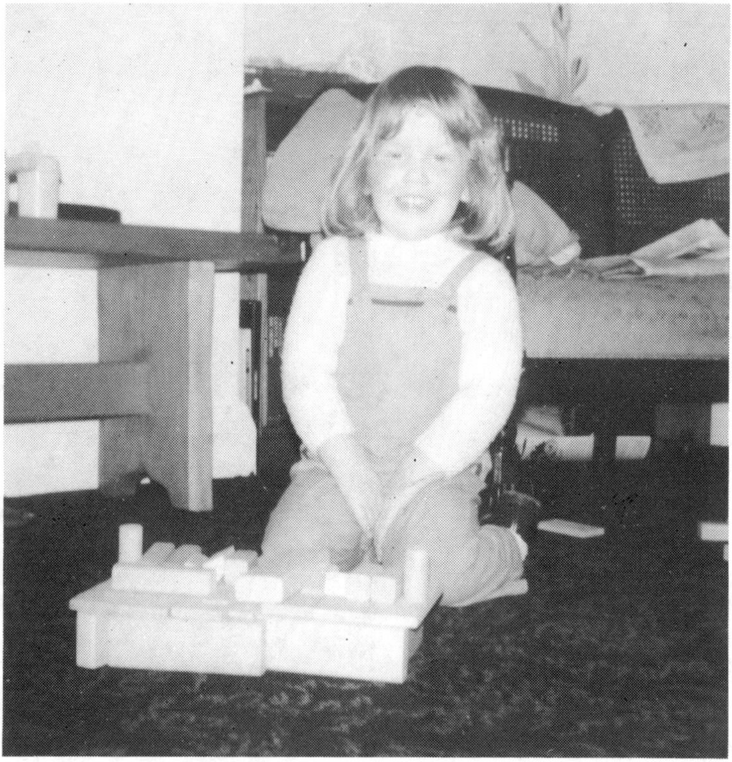